Sustained Simulation Performance 2016

Michael M. Resch · Wolfgang Bez
Erich Focht · Nisarg Patel
Hiroaki Kobayashi
Editors

Sustained Simulation Performance 2016

Proceedings of the Joint Workshop
on Sustained Simulation Performance,
University of Stuttgart (HLRS)
and Tohoku University, 2016

 Springer

Editors
Michael M. Resch
High Performance Computing Center
  (HLRS)
University of Stuttgart
Stuttgart
Germany

Wolfgang Bez
NEC High Performance Computing
  Europe GmbH
Düsseldorf
Germany

Erich Focht
NEC High Performance Computing
  Europe GmbH
Stuttgart
Germany

Nisarg Patel
High Performance Computing Center
  (HLRS)
University of Stuttgart
Stuttgart
Germany

Hiroaki Kobayashi
Cyberscience Center
Tohoku University
Sendai
Japan

Figure on Front Cover: Domain decomposition of a hierarchical Cartesian mesh. A Hilbert curve is used to partition the grid at a relatively coarse refinement level. Due to the depth-first ordering of the cells, this leads to complete subtrees being distributed among the available MPI ranks, improving the parallel performance of coupled multiphysics simulations

Figure on Back Cover: Hierarchical Cartesian mesh with local refinement towards the lower boundary. Among neighbouring cells, the level difference is at most one, leading to a size ratio of 2:1 (2D) or 4:1 (3D) between the cells

ISBN 978-3-319-83574-7   ISBN 978-3-319-46735-1   (eBook)
DOI 10.1007/978-3-319-46735-1

Mathematics Subject Classification (2010): 68Wxx, 68W10, 68Mxx, 68U20, 76-XX, 86A10, 70FXX, 92Cxx, 65-XX

Printed on acid-free paper

This Springer imprint is published by Springer Nature
The registered company is Springer International Publishing AG
The registered company address is: Gewerbestrasse 11, 6330 Cham, Switzerland

Preface

The field of high-performance computing is currently witnessing a significant shift of paradigm. Ever-larger raw number crunching capabilities of modern processors are in principle available to computational scientists. Imperative knowledge of efficiently exploiting modern processors and performance achievements in the scientific community is growing by leaps and bounds.

On the other hand, many areas of computational science have reached a saturation in terms of problem size. Scientists often do no longer wish to solve larger problems. Instead, they wish to solve smaller problems in a shorter time. The current architectures, however, are much more efficient for large problems than they are for the more relevant smaller problems.

This series of workshops focuses on *Sustained Simulation Performance*, i.e., high-performance computing for real-application use cases, rather than on-peak performance, which is the scope of artificial problem sizes. The series was established in 2004, initially named Teraflop Workshop, and renamed Workshop for Sustained Simulation Performance in 2012. In general terms, the scope of the workshop series has shifted from optimization for vector computers to emphasis on future challenges, productivity, and exploitation of current and future high-performance computing systems.

This book presents the combined results of the 22nd and 23rd installment of the series. The 22nd workshop was held at the High-Performance Computing Center, Stuttgart, Germany, in December 2015. The 23rd workshop was held in March 2016 at Sendai, Miyagi, Japan, and organized jointly with the University of Tohoku, Sendai, Japan.

The topics studied by the contributed papers include exploitation of HPC systems (Part I) and numerical computations and approach toward multi-physics applications (Part II).

We would like to thank all the contributors and organizers of this book and the sustained simulation performance project. We thank especially Prof. Hiroaki Kobayashi for the close collaboration over the past years and are looking forward to intensify our cooperation in the future.

Stuttgart, Germany   Michael M. Resch
August 2016   Nisarg Patel

Contents

Part I
Exploitation of Existing HPC Systems: Potentiality, Performance and Productivity

Parallel Algorithms: Theory, Practice and Education

Vl. V. Voevodin

Abstract Each new computing platform required software developers to analyze the algorithms over and over, each time having to answer the same two questions. Does the algorithm possess the necessary properties to meet the architectural requirements? How can the algorithm be converted so that the necessary properties can be easily reflected in parallel programs? *Changes in computer architecture do not change algorithms*, but this analysis had to be performed again and again when a program was ported from one generation of computers to another, largely repeating the work that had been done previously. Is it possible to do the analysis "once and for all," describing all of the key properties of an algorithm so that all of the necessary information can be gleaned from this description any time a new architecture appears? As simple as the question sounds, answering it raises a series of other non-trivial questions. Moreover, creating a complete description of an algorithm is not a challenge, it is a large series of challenges, and some of them are discussed in the paper.

1 Introduction

Parallel computing system architectures have gone through at least six generations over the past 40 years, each requiring its own algorithm properties and a special program writing style. In each case, it was important not only to find suitable features for the algorithms, but also to express them properly in the code, using special programming technologies. In fact, each new generation of computing architecture required a review of the entire software pool.

The generation of **vector pipeline computers** got off to a rapid start in the early seventies with the launch of the Cray-1 supercomputer. Machines of this class were based on pipeline processing of data vectors, supported by vector functional units and vector instructions in machine code. Full vectorization was the most efficient program implementation, which implied complete replacement of any innermost loops in the program body with vector instructions. Hence the requirements for algorithms and

Vl.V. Voevodin (✉)
Lomonosov Moscow State University, Moscow, Russia
e-mail: voevodin@parallel.ru

© Springer International Publishing AG 2016
M.M. Resch et al. (eds.), *Sustained Simulation Performance 2016*,
DOI 10.1007/978-3-319-46735-1_1

3

programs: parallelism was to be expressed in the form of independent iterations of the program's innermost loops. If that representation could be found and the innermost loops are vectored, the program would run efficiently.

During the eighties, computers came with not one, but several independent vector pipeline processors: **vector parallel computers**. The requirements for algorithm structure changed again. To support pipeline processing, inner loop parallelism was used as before. But this time, an additional parallelism resource was to be found within the algorithms that would support the independent operation of several processors. Inner loop parallelism was used to support vectorization, and outer loop parallelism to support the simultaneous operation of several CPUs.

The next generation to become commonplace during the nineties were **massively parallel distributed memory computers**, based on thousands of processors. Two actions were needed to make a program efficient. First, a substantial parallelism resource had to be identified in an algorithm to ensure the independent operation of many processors. Second, it was also important to distribute data between computing nodes to minimize data exchange during the course of program execution. This required not just another review of the algorithm pool based on the new programming technologies (with MPI becoming the de-facto standard), but also completely rewriting the software.

Shared memory computers also appeared actively. Shared memory substantially simplified the interaction between processors, making it easier to write parallel programs. Data distribution was no longer a major consideration as global address space and global shared variables eliminated many complex data handling issues. OpenMP technology appeared to reflect the new paradigm of parallel program operation. Shared memory computers required a new parallel program model, new means and methods of programming and new constructs which meant programs had to be rewritten yet again.

Computers combining the features of the two previous classes, **computing clusters with distributed memory, based on shared memory nodes**, appeared during the early 2000s. With these systems, one part of the parallelism resource inherent in an algorithm was to be kept for using a certain number of independent nodes, and the other for using several processors or cores within each node. In parallel applications, the first part was described through MPI and the second part through OpenMP. Converting an algorithm to efficiently use these features of the architecture was no trivial task, and was further complicated by the need to determine the proper data distribution for the MPI part.

About 8 years ago, **accelerators were first added** to the computer architecture— first as graphics processing units, and then as Xeon Phis. Now these devices can be found everywhere, including big clusters [2]. But what did the addition of accelerators mean for analyzing algorithm properties? It meant that a substantial parallelism resource needed to be identified in an algorithm for using many computing nodes. More parallelism needs to be present in each parallel process to utilize multiple computing cores per node. Moreover, enough parallelism needs to be left to use the accelerator features. Computers, like the need for parallelism, had become heterogeneous, which required revising the algorithm properties once more.

2   What is a Complete Description of the Algorithm Properties?

Each new computing platform required software developers to analyze the algorithms over and over, each time having to answer the same two questions. Does the algorithm possess the necessary properties to meet the architectural requirements? How can the algorithm be converted so that the necessary properties can be easily reflected in parallel programs? *Changes in computer architecture do not change algorithms*, but this analysis had to be performed again and again when a program was ported from one generation of computers to another, largely repeating the work that had been done previously.

This begs a natural question: is it possible to do the analysis "once and for all," describing all of the key properties of an algorithm so that all of the necessary information can be gleaned from this description any time a new architecture appears? As simple as the question sounds, answering it raises a series of other questions. What does it mean "to perform analysis" and what exactly needs to be studied? What kind of "key" properties need to be found in algorithms to ensure their efficient implementation in the future? What form can (or should) the analysis results take? What makes a description of algorithm properties "complete?" How does one guarantee that a description is complete and that all of the relevant information for any computer architecture is included?

The questions are indeed numerous and non-trivial. Obviously, a complete description needs to reflect many ideas: computational kernels, determinacy, information graphs, communication profiles, a mathematical description of the algorithm, performance, efficiency, computational intensity, the parallelism resource, serial complexity, parallel complexity…[3] All of these concepts, and many others, are used to describe an algorithm's properties from different perspectives, and they all are quite necessary in practice under various situations.

To immediately introduce some order to these diverse concepts, one can begin by breaking up an algorithm's description into two parts. The first part is dedicated to the algorithm's theoretical properties, and the second part describes its particular implementation features. This division allows the machine-independent properties of algorithms to be separated from the numerous issues arising in practice. Both parts of the description are equally important: the first one describes the algorithm's theoretical potential, and the second one demonstrates the practical use of that potential. The first part of the description explains the mathematical formulation of the algorithm, its computational kernel, input and output data, information structure, parallelism resources and properties, determinacy and computational balance of the algorithm, etc. The second part contains information on an algorithm's implementation: locality, performance, efficiency, scalability, communication profile, implementation features on various architectures, and so on.

3   Why Is It Hard to Describe Algorithms?

Many of the ideas described above are very well known. However, as you start describing the properties of real algorithms, you realize that creating a complete description of an algorithm is not a challenge, it is a large series of challenges! Unexpected problems arise at each step, and a seemingly simple action becomes a stumbling block. Let's look at the information structure of an algorithm mentioned above. It is an exceptionally useful term that contains a lot of information about the algorithm. An information graph is a convenient representation of an algorithm's information structure. In many cases, looking at the information graph is enough to understand its parallel implementation strategy. Figure 1a, b show the information structure for typical computational kernels in many algorithms, Fig. 1c shows the information structure of a Cholesky decomposition algorithm.

An information graph can be simple for many examples. However, in general, the task of presenting an information graph is not a trivial exercise. To begin with, a graph can potentially be infinite, as the number of vertices and arcs is determined by the values of external input variables which can be very large. In this situation it helps to look at likenesses: graphs for different values of external variables look very "similar" to one another, so it is almost always enough to present one small graph, stating that the graphs for other values will look "exactly the same." Not everything is so simple in practice, however; and one should be very careful here.

Next, an information graph is potentially a multi-dimensional object. The most natural coordinate system for placing vertices and arcs in an information graph relies

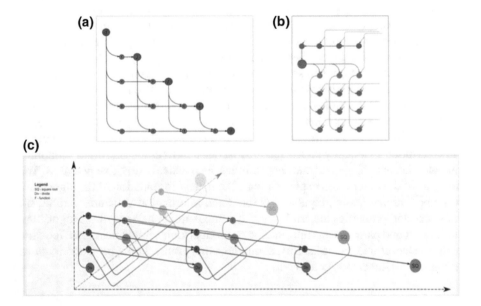

Fig. 1 Information structure of various algorithms

on the nested loops in an algorithm's implementation. If nested loops do not go deeper than three levels (as in a classical matrix multiplication algorithm), the graph can be placed in the traditional three-dimensional space. More complex looping constructs with 4 or more nesting levels require special methods for presenting and displaying the graph. But even if the number of dimensions does not exceed three, how does one make the graphical presentation informative? Figure 2a shows a graph in its entirety, which is barely comprehensible. Various projections of the same graph are shown in Figs. 2b, d, which can help assess an algorithm's parallelism potential, but these aren't always helpful...

Related questions also arise: how does one visualize the parallelism potential and illustrate parallel implementation methods for the algorithm? Sometimes a canonical parallel layer form [3] comes in handy, which reflects both the algorithm's parallel complexity and the fastest method for its parallel implementation (within the infinite parallelism concept), but it is very difficult to build and not always feasible. Figure 3 shows the sequential execution of the fragment in Fig. 2a in five steps. Red indicates vertices within the current level that can be executed in parallel on the current step. Green indicates vertices executed in previous steps, and white indicates vertices that can only be executed in subsequent steps. By visualizing the step-by-step sequential movement of the red vertices, one can evaluate the parallelism available at each step. How does one find, analyze, describe and display the canonical parallel layer form? The question remains open for arbitrary programs.

The issues of data locality and computation locality are of paramount importance in describing an algorithm's properties and its implementations. Locality is

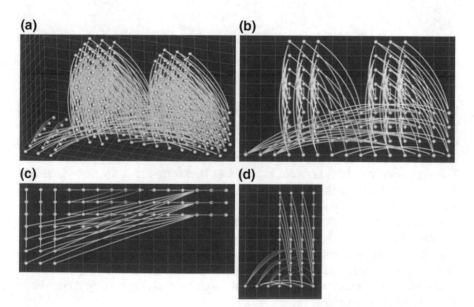

Fig. 2 Methods of displaying an algorithm's information structure

(1)   **(2)**   **(3)**

(4)   **(5)**

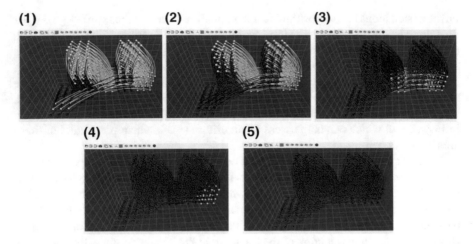

Fig. 3 Sequence of steps in the parallel execution of an algorithm based on a canonical parallel layer form

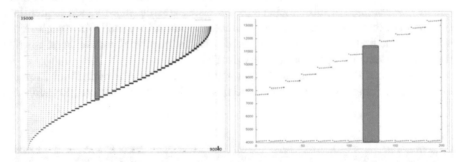

Fig. 4 A description of data locality in programs using memory access profiles

what determines program execution efficiency on modern computing platforms. To get the complete picture of an algorithm's particular implementation features, it is important to analyze both temporal and spatial locality, noting positive and negative factors related to locality, and the conditions and situations by which they are caused. However, even a quick look makes it obvious that there are many more questions than answers in this area. What methodology can be used to evaluate the temporal and spatial data locality in the programs? How can one compare temporal and spatial data locality between different programs? Figure 4 shows the memory access profiles for two programs, indicating the memory address after each memory access operation. Which program has better temporal and/or spatial data locality? In some cases, memory usage templates help: they are simple and their characteristics are predetermined, but once again, the issue of carefully studying locality properties in an arbitrary program generally remains open.

Another interesting question is related to how data locality is related to the algorithm structure. In other words, can we predict data locality in a given program by using just the information about its algorithm? On the one hand, there are no data structures in algorithms—they only appear in programs; so talking about locality for algorithms is not exactly right. On the other hand, it is the algorithm that determines the structure and properties of a program to be coded, including its locality. Many have probably heard the expression "the algorithm's locality" or "this algorithm has better locality than the other." How appropriate are these statements, given that algorithms do not contain data structures?

Determinacy is an important practical aspect of algorithms and programs, but how can one describe all of the potential sources which violate this property? A serious cause of indeterminacy in parallel programs is related to changes in the order of executing associative operations. A typical example is the use of global operations in Message Passing Interface (MPI) by a number of parallel processes, e.g., when summing the elements of a distributed array. The MPI runtime system chooses the order of execution on its own, assuming compliance with the associative law, which results in various round-off errors and ultimately in different results when executing the same application. This is a serious issue often encountered in massively parallel computing systems that causes results of parallel program execution to not be reproducible. If the analysis of an algorithm's structure shows that the resulting parallel application cannot work without global operations, this property must be included in the algorithm description. To analyze this problem properly, a communication profile should be built for the parallel program, pointing out the structure and interaction method between parallel processes. A clear definition of the communication profile hasn't been produced to date, so it is premature to consider in-depth analysis in this area.

Indeed, there are many open questions, and the list can go on. The main question that still remains unanswered is "What does it mean *to create a complete description of an algorithm?*" What must be included in this description, so that we can glean all of the necessary information from it every time a new computing platform appears? The task seems simple at first sight: an algorithm is just a sequence of mathematical formulas, often short and simple, which should easily be analyzed. But at the same time, no one can guarantee the completeness of such a description.

The properties of the algorithms and programs discussed in this work became the foundation for the AlgoWiki project [1]. The project's main goal is to provide a description for fundamental algorithm properties which will enable a more comprehensive understanding of their theoretical potential and their implementation features in various classes of parallel computing systems. The project is expected to result in the development of an open online encyclopedia based on wiki technologies which will be open to contributions by the entire academic and educational community. The first version of the encyclopedia is available at http://AlgoWiki-Project.org/en, where users can describe both their own pedagogical experience and their knowledge of specific parallel algorithms.

4 Conclusion

All of the issues discussed in this work are highly important for training future specialists [4–6]. Right from the beginning of the education process, focus should be placed on algorithm structure since it determines both the implementation quality and the potential for efficiently executing programs in a parallel environment. The algorithm structure and its close relationship to parallel computing system architecture are central ideas in parallel computing, which are included in many courses for Bachelor's and Master's degree programs at the Faculty of Computational Mathematics and Cybernetics at Lomonosov Moscow State University, as well as in the lectures and practical courses offered by the annual MSU Summer Supercomputing Academy [7]. We are also trying to expand this concept to the Supercomputing Consortium of Russian Universities [8] in order to develop a comprehensive supercomputer education system, rather than offering occasional training aimed at rectifying the situation.

Acknowledgements This project is being conducted at Moscow State University with financial support from the Russian Science Foundation, Agreement No 14-11-00190.

References

1. Antonov, A., Voevodin, V., Dongarra, J.: AlgoWiki: an open encyclopedia of parallel algorithmic features. Supercomput. Front. Innov. **2**(1), 4–18 (2015)
2. Dongarra, J., Beckman, P., Moore, T., Aerts, P., Aloisio, G., Andre, J.C., Barkai, D., Berthou, J.Y., Boku, T., Braunschweig, B., et al.: The international exascale software project roadmap. Int. J. High Perform. Comput. Appl. **25**(1), 3–60 (2011)
3. Voevodin, V.V., Voevodin. Vl.V.: Parallel Computing. BHV-Petersburg, St. Petersburg (2002). (in russian)
4. Computing Curricula Computer Science. http://ai.stanford.edu/users/sahami/CS2013 (2013)
5. Future Directions in CSE Education and Research, Workshop Sponsored by the Society for Industrial and Applied Mathematics (SIAM) and the European Exascale Software Initiative (EESI-2), http://wiki.siam.org/siag-cse/images/siag-cse/f/ff/CSE-report-draft-Mar2015.pdf (2015)
6. NSF/IEEE-TCPP Curriculum Initiative on Parallel and Distributed Computing. http://www.cs.gsu.edu/~tcpp/curriculum/
7. Summer Supercomputing Academy. http://academy.hpc-russia.ru/
8. Supercomputing Education in Russia, Supercomputing Consortium of the Russian Universities. http://hpc.msu.ru/files/HPC-Education-in-Russia.pdf (2012)

High Performance Computing and High Performance Data Analytics—What is the Missing Link?

Bastian Koller, Michael Gienger and Michael M. Resch

Abstract Within this book chapter, technologies for data mining, data processing and data interpreting are introduced, evaluated and compared. Especially, traditional High Performance Computing, and the newly emerging fields High Performance Data Analytics and Cognitive Computing are put into context in order to understand their strengths and weaknesses. However, the technologies have not been evaluated solely, but also the missing links between them have been identified and described.

1 Introduction

At this point of time, there are various technologies in the market that target data analysis, data processing, data interpreting and data mining. So far, it has not been clear if all of those technologies are direct competitors or can be seen in a complementary fashion. This book chapter therefore analyses the technologies carefully and introduces as well as compares their direct angles. Being more concrete, traditional High Performance Computing, the newly emerging field High Performance Data Analytics as well as Cognitive Computing are evaluated. In particular, the interactions between those technological fields are visualized in addition.

The book chapter is organized as follows: Section 2 is providing the High Performance Computing context, Sect. 3 is introducing High Performance Data Analytics whereas Sect. 4 compares the approaches and describes the missing links. Finally, Sect. 5 concludes this book chapter.

B. Koller (✉) · M. Gienger · M. Resch
High Performance Computing Center Stuttgart, Nobelstrasse 19,
70569 Stuttgart, Germany
e-mail: koller@hlrs.de

M. Gienger
e-mail: gienger@hlrs.de

M. Resch
e-mail: resch@hlrs.de

© Springer International Publishing AG 2016
M.M. Resch et al. (eds.), *Sustained Simulation Performance 2016*,
DOI 10.1007/978-3-319-46735-1_2

2   The Evolution of High Performance Computing

Within this section of the book chapter, a generic view on High Performance Computing (HPC) and its evolution over time is given. Although the purpose of such HPC systems is in principle the same, the available performance, the customer base as well as the computational and applications models changed in the last decade. In summary, various application areas such as computational fluid dynamics, climate or physics simulations are considered HPC relevant at the moment, which are executed on innovative systems that may be equipped by vector central processing units, by commonly used x86 processors or even accelerators.

2.1   Traditional High Performance Computing

High Performance Computing has been traditionally designed to solve problems that are too large and complex for common desktop computers or even workstations. Those systems enable a maximum of performance for memory, compute, storage or input/output (I/O) intensive applications and operations. However, with respect to their special design and the corresponding drastic costs, they clearly lack on the flexibility to combine all requirements into a unique general-purpose system.

Although there are self-appointed general-purpose systems in the worldwide HPC market, there is always a key application that drives the selection of such systems. Applications that require solely a high computational demand will result in a system architecture that is based on accelerators, whereas applications that require thousands of memory operations per second will rather tend to the vector or x86 architecture. Thus, due to the main area of applications and the corresponding costs, a HPC system is always tailored to its common applications so that "real" general-purpose systems cannot be seen in the markets.

2.2   Evolution Over Time

Within the last decade, there was a huge evolution with regards to the HPC systems. Reaching from vector machines to the widely adopted x86 architecture and modern accelerators, especially hardware evolved quickly. In the meantime, HPC systems with more than 1.000.000 cores are not an utopia any more[1] so that besides the efficiency of the systems, also the models and applications can benefit from the huge amount of provided computational performance.

But not just the hardware evolved, also the customer basis changes: industrial applications from the automotive world, academic applications dealing with, for instance, climate simulation as well as applications from small and medium sized

---

[1]Top500: http://www.top500.org

enterprises from various kinds of areas are targeting the High Performance Computing systems. However, with the evolved systems and the immense performance, also the execution models get more complicated. On the one hand, there are still traditional applications that require a huge amount of resources for a single run and on the other, parametric studies with less constant performance requirements but generating a huge amount of results are common in state-of-the-art HPC systems.

Nevertheless, HPC driving applications are still usual in the High Performance Computing area, but due to the changing application and executions models, general-purpose systems are becoming more evident as large computational intensive applications typically produce a huge amount of results. So there is currently a trade-off between providing generic systems that are flexible enough to cope with different kinds of workloads and such systems that are solely made to provide one single key performance type.

3 Towards High Performance Data Analytics

In contrast to Sect. 2, this chapter focuses High Performance Data Analytics (HPDA), a new emerging field for the High Performance Computing sector. High Performance Data Analytics target the efficient analytics of various kinds of data, reaching from structured up to unstructured as well as streaming data, which cannot be analysed anymore on standard workstations or Clouds due to their volume, their variety or their velocity.

3.1 Where Is It Needed?

As already highlighted in the introduction of this section, High Performance Data Analytics target the analysis of available (e.g. stored) or real-time streaming data. In contrast to HPC applications, HPDA requires typically not an extraordinary huge amount of compute performance, but rather a very broad I/O backend that is able to transfer data quickly enough to the actual processing engines.

The applications that typically cause such data intensive workloads are settled in the sensor technologies area, such as the evolving Internet of Things, the aligned Industry 4.0 and the cyber physical systems area. The physical sensors produce a huge amount of data that has to be analysed in time, sometimes even real-time to provide the corresponding actions. However, not just those industrial areas require the implicitly described system architecture, but also modern Internet stores with their designed customer marketing require as much as knowledge possible about their customers. This fact results in strong correlations of data that have to be analysed on huge-scale systems, since Clouds are not performing enough. Finally, not only the described applications require HPDA functionality, fine-grained models and their corresponding applications produce Terabytes of data in the meanwhile that cannot be analysed on a state-of-the-art HPC system anymore.

3.2   HPDA Concepts and Technologies

As already highlighted in the sections before, HPC and HPDA approaches in terms of hardware and software require different technologies. Therefore, these requirements will be discussed and addressed in particular in this sub-section to bridge the gap between both technologies.

In terms of hardware, data intensive workloads require different key performance indicators than standard HPC applications. The differences between both approaches are highlighted below:

- **Processors**
  In traditional HPC systems, fast processors with fast memory pipelines are focused. For HPDA systems, the amount of Floating Point Operations Per Second is still important, however the performance of the system is determined by the storage system.
- **Memory**
  The more memory available for data analytics, the better for the overall application execution since most of the data and results can be kept in memory instead of check pointing them to the storage backend. For HPC systems, the same statement holds, although much smaller memory systems are targeted than in the HPDA area.
- **Networks**
  Whenever data needs to be transferred, fast interconnects come into play. So both, HPC and HPDA systems require fast memory and latency-oriented networks in order to transfer the data efficiently.
- **Storage**
  Typical HPC systems provide a central system storage from which all the required data gets read and written. An approach like this is not possible for HPDA since the data accessibility is the key performance indicator for the whole applications. Therefore, data analytics systems provide fast local disks that can be used to provide and cache the data in order to optimize the application execution.

As can be seen, the main differences between HPC and HPDA systems are located in the area of processors and storages, since fast number-crunching processors are required for HPC only. In contrast, very fast input/output systems with large capacity are mandatory for efficient data processing.

The software requirements come along with the hardware requirements. In contrast to traditional HPC applications, which require programming models and paradigms such as message passing or shared memory parallelism, data analytics applications rely on in-memory processing and programming languages such as Java, Python or Scala. So the most important applications for data analytics are currently the Apache tools Spark[2], Hadoop[3], Storm[4] and Flink[5] as well as some smaller projects such as Disco Project[6], DataTorrent[7] or BashReduce[8].

Most of those applications build on the MapReduce algorithm, which has been introduced by the global player Google[9]. The MapReduce algorithm consists of three phases—map, shuffle and reduce, whereas the map and the reduce parts are directly specified by the user in order to allow parallel processing of data on manifold machines. Using his concept enables processing different kinds of data, reaching from structured data including files and databases up to unstructured and real-time data such as online data composed of several data structures.

3.3 A Practical Application Making Use of HPC and HPDA

In order to proof the statements of the last sections and sub-sections, the information shall be complemented with a practical example from the Global Systems Science community, which represents an emerging field in the HPC sector. Within the EC-funded CoeGSS project[10], a set of applications is focused that require particular workflows to retrieve the results. In particular, the workflow foresees HPDA, huge-scale HPC, small-scale HPC and visualization to generate synthetic populations, execute the resulting agent-based models and finally, visualize the results [1]. For clarification, the workflow and its targeted technologies is depicted in Fig. 1.

Thus, those kinds of applications demonstrate that there is a new need to support other methods and techniques than the classical HPC applications demand. As a consequence, being competitive in terms of hardware and software reaches a new level of complexity.

4 The Missing Link

Summarizing the previously mentioned evolution scenarios for High Performance Computing and the raise of High Performance Data Analytics, this seems as a promising and valuable way to go. However the deeper one dives into the implications of the use of these technologies and the potential they provide, it becomes obvious that

---

[2]Apache Spark: http://spark.apache.org

[3]Apache Hadoop: http://hadoop.apache.org

[4]Apache Storm: http://storm.apache.org

[5]Apache Flink: http://flink.apache.org

[6]DiscoProject: http://www.discoproject.org

[7]DataTorrent RTS: https://www.datatorrent.com

[8]BashReduce: https://github.com/erikfrey/bashreduce

[9]Google Inc.: http://www.google.com

[10]Centre of excellence for Global Systems Science: http://www.coegss.eu

Fig. 1 CoeGSS application
workflow

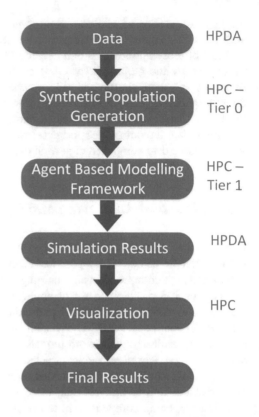

the resulting outputs, especially in terms of data variety and data size get hard to handle for a human in the loop.

We see a tendency in so-called "business-ready solutions" to stress the support of the human in the loop by application of technological fields such as machine learning, artificial intelligence or cognitive computing. For the remainder of this paper we will stick to the term cognitive computing as a placeholder for the above mentioned disciplines, which can be described as the variety of scientific disciplines of Artificial Intelligence and Signal Processing[11]. A similar view has been presented by James Kobielus, Big Data Evangelist, 2013, in a blog entry on *Cognitive Computing: Relevant at all Speeds, Scales and Scopes of Thought*, where he defines cognitive computing as

> the ability of automated systems to handle the conscious, critical, logical, attentive, reasoning mode of thought that humans engage in when they, say, play Jeopardy or try to master some academic discipline.

---

[11]Wikipedia Definition of Cognitive Computing: https://en.wikipedia.org/wiki/Cognitive_computing

4.1 Cognitive Computing

The principles of cognitive computing are not new, and nearly everyone who is in the Information Technology business has at a certain point in time heard of this topic. Thus is it also not surprising, that it's base assumptions and ideas were even reported already at the end of the 19th century, when Boole proposed its book on "The Laws of Thoughts" [2]. Even though this was just a conceptual approach, and the first programmable computer by Zuse needed As already mentioned before, during the evolution of these principles, the domain of cognitive methodologies and artificial intelligence went either side by side or showing clear overlaps. A variety of theories and implementation approaches were taken, the probably most prominent ones being so far IBM's Watson [3] and the recently presented AlphaGo [4].

4.2 Benefits

Figure 2 shows how High Performance Computing, High Performance Data Analytics and Cognitive Techniques can complement each other. High Performance Computing (HPC) delivers the needed processing power for those kind of applications, requiring massive parallel execution. At the same time, these kind of applications produce partially enormous amounts of data, which may be too big to be manually analysed, even having current support tools at hand. Thus the discipline of High Performance Data Analytics can be used to analyse and handle these (and other sources' data sets) in a sufficient way. Cognitive techniques can provide support to both disciplines, to help to interpret and present the results in a best possible way.

In a general way, the expected benefits from applying these concepts, are manifold. In general support for those fields where big amounts of data are collected, handled and interpreted is improved, examples are:

- Enhanced analysis of business potentials of new offerings/new activities. This can reach from the virtual testing of new opportunities, e.g. in drug design or on combined virtual and real world simulations such as finding new geographic locations for drilling
- Support of staff (e.g. engineers) in decision processes by providing them a selection of potential paths to follow
- Improving Operations by understanding of performed operations and their parameters, so that either in real time or after longer-duration analysis processes can be optimized

Taken this complementarity into account, the workflow as described in Fig. 1 can be extended to the one presented in Fig. 3.

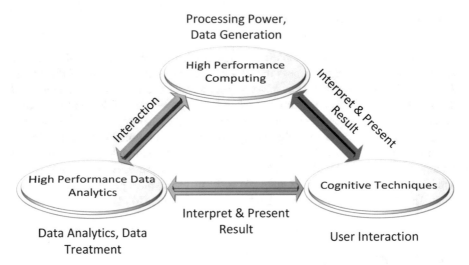

Processing Power,
Data Generation

Fig. 2 Cognitive Techniques complementing the global picture of HPC and HPDA

Fig. 3 Extending the GSS
workflow with cognitive
techniques support

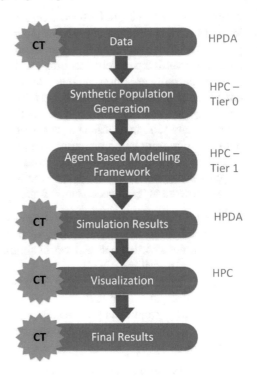

4.3   Available Technologies

Within this document, we also want to have a short look at those technologies, which may act as baseline to realize an integration of cognitive concepts into a traditional HPC/HPDA based workflow (e.g. the one presented in Fig. 3.

In the case of Watson, a variety of APIs is available for selected developers and business users, as well as the Watson Analytics Solution[12]. Furthermore there is a variety of Open Source alternative available, which shall be discussed on a high level in the following overview:

DARPA DeepDive

DeepDive [5, 6] is a free version of a Watson like system. It was developed within the frame of the US Defense Advanced Research Projects Agency (DARPA) and in opposite to Watson has the aim to extract structured data from unstructured data sources. DeepDive uses machine learning technologies to train itself and targets especially those users with moderate to no machine learning expertise.

UIMA

Apache Unstructured Information Management (UIMA)[13] is supporting the analysis of large sets of unstructured information. Its an implementation of the Oasis Unstructured Information Management standard[14] **OpenCog**

OpenCog [7] is a project targeting artificial intelligence and delivering an open source framework. One output of OpenCog is the cognitive architecture OpenCog Prime [8] for robot and virtual embodied cognition.

5   Conclusions

The previous sections have pointed out that High Performance Computing and High Performance Data Analytics can be seen as rather complementary approaches, then as direct competitors. Even though there are activities to provide a common software stack, which may run on both, HPC and HPDA specific hardware, there is only a subset of concrete problems in the problem space which can be addressed efficiently in such a manner. Mainly, this is a result of the partially quite different hardware set up of the respective technological environment.

Now, assuming that HPC and HPDA work with a high performance, we also have to face the fact that the size and amount of data sets proceeded and again resulting from this processing enter a dimension, which makes a satisfactory manual processing by a human in the loop (e.g. an engineer) nearly impossible. Thus we see that even if there is an issue (e.g. data analytics) solved with those appliances, another issue pops up which is the understanding and respectively handling of information.

---

[12]http://www.predictiveanalyticstoday.com/ibm-watson-analytics-beta-open-business/

[13]http://uima.apache.org/

[14]https://www.oasis-open.org/committees/download.php/28492/uima-spec-wd-05.pdf

For that purpose we have introduced cognitive technologies, which can act as some sort of "helper" technology to simplify the life of the end user and enable for improved use of simulation results. This technology, even if it appears to be still in its infancy, can support the (human) end user and provide decision baselines allowing improved processing of information. We have shown that a variety of implementations already exist, next steps need to see in how far they can cover the requirements of selected use cases.

References

1. Wolf, S., Paolotti, D., Tizzoni, M., Edwards, M., Fuerst, S., Geiges, A., Ireland, A., Schuetze, F., Steudle, G.: D4.1 - First report on pilot requirements. http://coegss.eu/wp-content/uploads/2016/03/CoeGSS_D4_1.pdf
2. Boole, G.: Investigation of the Laws of Thought on Which are Founded the Mathematical Theories of Logic and Probabilities (1853)
3. Ferrucci, D.A., Brown, E.W., Chu-Carroll, J., Fan, J., Gondek, D., Kalyanpur, A., Lally, A., Murdock, J.W., Nyberg, E., Prager, J.M., Schlaefer, N., Welty, C.A.: Building Watson: an overview of the DeepQA project. AI Mag. **31**(3), 59–79 (2010)
4. Silver, D., Hassabis, D.: AlphaGo: mastering the ancient game of Go with Machine Learning, Blogpost. https://research.googleblog.com/2016/01/alphago-mastering-ancient-game-of-go.html (2016)
5. Niu, F., Zhang, C., Re, C., Shavlik, J.W.: DeepDive: web-scale knowledge-base construction using statistical learning and inference. 884. In: VLDS: CEUR-WS.org. (CEUR Workshop Proceedings), pp. 25–28 (2012)
6. Zhang, C.: DeepDive: a data management system for automatic knowledge base construction, Ph.D. Dissertation, University of Wisconsin-Madison (2015)
7. Hart, D., Goertzel, B.: OpenCog: a software framework for integrative artificial general intelligence. In: Wang, P., Goertzel, B., Franklin, S. (eds.) 'AGI', pp. 468–472. IOS Press (2008)
8. Goertzel, B.: OpenCog Prime: a cognitive synerfy based architecture for artificial general intelligence
9. Hurwitz, J.S., Kaufman, M., Bowles, A.: Cognitive Computing and Big Data Analytics. Wiley, Indianapolis (2015)

A Use Case of a Code Transformation Rule Generator for Data Layout Optimization

Hiroyuki Takizawa, Takeshi Yamada, Shoichi Hirasawa and Reiji Suda

Abstract Xevolver is a code transformation framework for users to define their own code transformation rules. In the framework, an abstract syntax tree (AST) of an application code is written in an XML format, and its transformation rules are expressed in the XSLT format, which is a standard XML format to describe XML data conversion; an AST and its transformation rules are both written in XML. Since it is too low-level for standard users to manually write XSLT rules, Xevtgen is now being developed as a tool to generate such rules from simple code description. In Xevtgen, users basically write just two code patterns, the original and transformed code patterns. Then, Xevtgen automatically generates a transformation rule that transforms the original code pattern to the transformed one. The generated rule is written in XSLT, and hence usable by other tools of the Xevolver framework. This article shows a use case of using Xevtgen for data layout optimization, and discusses the benefits of using the tool.

1 Introduction

When data are stored in a memory space, the layout of data often needs to be optimized so as to make a better use of memory hierarchy and architectural features. Today, such data layout optimization is critically important to achieve high performance on a modern high-performance computing (HPC) system, because the system

H. Takizawa (✉) · T. Yamada · S. Hirasawa
Graduate School of Information Sciences, Tohoku University, Sendai, Japan
e-mail: takizawa@tohoku.ac.jp

T. Yamada
e-mail: tyamada@sc.cc.tohoku.ac.jp

S. Hirasawa
e-mail: hirasawa@sc.cc.tohoku.ac.jp

R. Suda
Graduate School of Information Science and Technology,
The University of Tokyo, Tokyo, Japan
e-mail: reiji@is.s.u-tokyo.ac.jp

© Springer International Publishing AG 2016
M.M. Resch et al. (eds.), *Sustained Simulation Performance 2016*,
DOI 10.1007/978-3-319-46735-1_3

21

performance is very sensitive to memory access patterns. Memory access can easily become a performance bottleneck of an HPC application.

The data layout of an application can be optimized by changing data structures used in the code. One problem is that a human-friendly, easily-understandable data representation is often different from a computer-friendly data layout. This means that, if the data layout of a code is completely optimized for computers, the code may be no longer human-friendly.

We have been developing a code transformation framework, Xevolver, so that users can define their own rules to transform an application code [1, 2]. In this article, such a user-defined code transformation rule is adopted to separate the data representation in an application code from the actual data layout in a memory space. Instead of simply modifying a code for data layout optimization, the original code is usually maintained in a human-friendly way and then mechanically transformed just before the compilation so as to make the transformed code computer-friendly.

One important question is how to describe code transformation rules. A conventional way of developing such a code translator is to use compiler tools, such as ROSE [3]. Actually, at the lowest abstraction level, Xevolver allows users to describe a code transformation rule as an AST transformation rule. Since AST transformation is exactly what compilers internally do, compiler experts can implement various code transformation rules by using the framework. However, standard programmers who optimize HPC application codes are not necessarily familiar with such compiler technologies. Therefore, we are also developing several high-level tools to describe the rules more easily.

Xevtgen [4] is one of high-level tools to help users define custom code transformation rules. This article shows a use case of Xevtgen for data layout optimization, and discusses how it can help users define their own transformations.

2 Data Layout Optimization

In many cases, an HPC application code is written in a low-level programming language such as C/C++ and Fortran. In such a language, a data structure mostly corresponds to a specific data layout. In practice, thus, the data layout of an HPC application is usually altered by changing the data structure in the code.

A typical example of data layout optimization is so-called AoS-to-SoA conversion [5]. Generally, an array of structures (AoS) is likely to be human-friendly, leading to high code maintainability and readability. For example, the following C code defines an AoS data structure, in which each point is a pair of two variables, x and y.

```
struct { double x, y; } point2d[N];
```

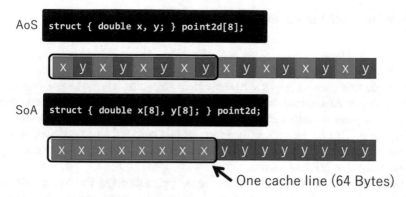

Fig. 1 Data layout in a memory space

This would be an intuitive representation of points in a 2-dimensional space. On the other hand, a structure of arrays (SoA) often leads to better memory access patterns. The following is an example of SoA, in which x and y are both arrays.

```
struct { double x[N], y[N]; } point2d;
```

Each data structure leads to a different data layout in a memory space as shown in Fig. 1. In the case of AoS, x and y appear alternately in the memory space. On the other hand, in the case of SoA, each of the two arrays organizes a continuous memory region. For example, when x of every point is sequentially accessed, it is obvious that SoA has a higher spacial locality of reference. As a result, SoA can potentially use cache memory more effectively. In this way, the data layout in a memory space can significantly affect the performance of an HPC application.

One severe problem is that data layout optimization needs to modify many places in an application code, and degrades the code maintainability and readability. As long as the data representation in an application code corresponds to its actual data layout in a memory space, data layout optimization results in drastic code modification. Moreover, it could reduce the performance portability across different systems because different systems potentially prefer different data layouts. Programmers may need to maintain multiple versions of one application code, e.g., using many #ifdef's, especially if the application needs to be performance-portable. This approach likely makes the code unmaintainable.

One idea to solve this problem is to use code transformation. For example, various compiler tools are available to transform a code instead of directly modifying the code. However, transformation for data layout optimization is generally specific to a particular application code. This is because such a transformation rule usually depends on the definition of the data structure. In general, it is not affordable to use a compiler tool to develop a custom code translator for individual applications. Accordingly, we need an easier way to define a custom code transformation rule.

3 Code Transformation Rule Generation with Xevtgen

Suppose that a legacy code written in Fortran uses a human-friendly data structure.
Then, this article discusses how the code should be converted to another version
of the code that uses a computer-friendly data structure. Thus, the purpose of this
conversion is to develop a code using human-friendly data representation and execute
it using computer-friendly data layout. To this end, we transform the code instead of
modifying it. That is, the code is transformed just before the compilation, and then
the transformed version is passed to the compiler. As a result, application developers
maintain only the original version.

One difficulty is that this conversion needs application-specific code transformations in many cases. The Xevolver framework has been developed to allow standard
users to implement such an application-specific code transformation. First, Xevolver
internally converts an application code to its AST, represents the AST in an XML
data format, and then exposes it to users. The users can apply any transformations to
the exposed AST. After the AST transformation, Xevolver converts the transformed
AST to its corresponding code. At the lowest abstraction level, Xevolver internally
uses XSLT [6] to express an AST transformation rule, because XSLT is a standard
XML data format to describe XML data conversion. It should be noted that an AST
and its transformation rules are both written in XML, i.e., text data. In comparison
with other data formats, it would be easy to quickly develop a simple tool to generate
and/or process text data for application-specific purposes. Therefore, we have been
developing various tools for the Xevolver framework and discussing the practicality
and superiority of using user-defined code transformation for optimizing an HPC
application code.

Xevtgen [4] is one of the tools to generate XSLT rules for AST transformation.
In the case of using Xevtgen, users do not need to consider their code transformation rules at an AST level. Instead, they write two code patterns, the original and
transformed code patterns. Then, Xevtgen automatically generates an XSLT rule of
custom code transformation that transforms the original code pattern to the transformed one.

Figure 2 shows an example of "dummy" Fortran codes to express custom code
transformations. Such a dummy code is used by Xevtgen to generate XSLT rules for
AST transformation. In Fig. 2, the original and transformed code patterns are written
using !$xev tgen src and !$xev tgen dst directives, respectively. In the
dummy code, some special variables can be used to express a code pattern. For
example, a variable, idx, defined in Line 27 matches any expression, and hence is
used in the rule to indicate that an array index can be any expression. If such a variable
appears in both of the original and transformed code patterns, the corresponding part
of the code pattern is copied from the original code to the transformed one. Similarly,
a variable, cmp, matches any name so that a component of aos_t is translated to
its corresponding component of soa_t.

Reading a dummy Fortran code with special directives as in Fig. 2, Xevtgen generates XSLT rules that implement code transformations expressed in the dummy

```
1   !!!!!!!!!!!!!!!!!!!!!!!!!!!!!!!!!!!!!!!!!!!!!!!!!!!!!!!!!!!!!!
2   !! A SIMPLE EXAMPLE OF DUMMY FORTRAN CODE                  !!
3   !!!!!!!!!!!!!!!!!!!!!!!!!!!!!!!!!!!!!!!!!!!!!!!!!!!!!!!!!!!!!!
4   program dl
5
6     !! Rule 1
7     !! Replace the type definition of AoS data structure
8     !$xev tgen src begin
9       type aos_t
10            real:: x
11            real:: y
12      end type aos_t
13
14      type(aos_t) aosdata(100)
15    !$xev end tgen src
16    !$xev tgen dst begin
17      type soa_t
18            real:: x(100)
19            real:: y(100)
20      end type soa_t
21
22      type(soa_t) soadata
23    !$xev end tgen dst
24
25    !! Rule 2
26    !! Change the way of accessing the data structure
27    !$xev tgen var(idx) exp
28    !$xev tgen var(x) name
29    !$xev tgen trans exp src('aosdata(idx)%x') dst('soadata%x(idx)')
30
31    end program dl
```

Fig. 2 A dummy Fortran code written for Xevtgen

code. This is the most important feature of Xevtgen. In the case of using Xevtgen, users do not need any knowledge about AST transformation; they can write a dummy code to express custom code transformations if they know Fortran programming and several !$xev tgen directives. Using Xevtgen, users can easily and quickly define a code transformation dedicated to a particular application code. In practice, such an application-specific code transformation rule does not need to have high generality if necessary rules can easily be defined for each application code. As a result, an application code can remain human-friendly from the viewpoint of application programmers, and user-defined code transformations are applied to the code just before the compilation so that the transformed code becomes computer-friendly and/or compiler-friendly.

Generally, two kinds of code transformations are required for AoS-to-SoA conversion. One is called a *type definition transformation rule* that replaces the type definition of an AoS data structure to that of an SoA data structure. An example of a type definition transformation rule is Rule 1 in Fig. 2. The other is called a *variable reference transformation rule* that changes the way of accessing the data structure. An example of a variable reference transformation rule is Rule 2 in Fig. 2.

In many cases, a type definition transformation rule is explicitly defined by users as in Fig. 2. It often looks like simple textual replacement. Since a type definition appears only once in an application code, simple textual replacement is usually an easy and appropriate way to achieve type definition transformation.

On the other hand, a variable reference transformation rule often needs to be expressed using generalized code patterns rather than concrete texts. This is because a data structure could have various components. If simple text replacement is adopted for this kind of transformation, the rule will be very wordy. To make matters worse, since those variable references could be scattered over a whole application code, it is a tedious and error-prone task to manually change all of them for using a different data structure. If Xevtgen is used and there is a general code pattern that matches all the variable references, a variable reference transformation rule could be simple as shown in Fig. 2. Once such a simple rule is defined, all the variable references are mechanically transformed based on the rule.

In the dummy code in Fig. 2, the part of a directive surrounded by back quotes, such as `aosdata(idx)%x`, is a Fortran expression. The expression must be valid when Xevtgen translates the dummy code to XSLT rules. Hence, a Tgen variable, x, is defined to have the same variable name as a component of aos_t so that `aosdata(idx)%x` is a valid expression. Since x is a Tgen variable as well as a component of aos_t, it matches any name when the generated XSLT rule is applied to a Fortran code. As a result, the generated rule replaces not only aos(i,j,k)%x but also aos(i,j,k)%y at the AoS-to-SoA conversion.

4   Discussions

In this article, Xevtgen is used for AoS-to-SoA conversion to discuss how Xevtgen helps users define application-specific code transformations. In the following evaluation, the classic static memory allocation version of the Himeno benckmark [7] written in Fortran77 is first modified by hand so as to use AoS data representation, and then a dummy Fortran code for Xevtgen is written to generate XSLT rules for AoS-to-SoA conversion. In addition, the performance difference between the AoS version and the SoA version is also shown to discuss how important data layout optimization is for modern computing systems.

Figure 3 shows a dummy Fortran code used by Xevtgen for AoS-to-SoA conversion in this use case.

For variable reference transformation, the rule in Fig. 3 can be expressed as a simple code of only four lines because there is a clear pattern in the transformation. The code pattern of accessing the original AoS data structure is mechanically replaced with the code pattern of accessing arrays. Since variable references are scattered over a whole application code, it is significantly helpful if all of them are transformed based on a certain transformation rule, especially in the case of optimizing the data layout of a large-scale practical application code. Even in a relatively-small code of the Himeno benchmark, 61 variable references need to be modified for changing the data structure.

```
1    program dl_himeno
2      !! Rule 1
3      !! Replace the type definition of AoS data structure
4      !$xev tgen src begin
5      type aos_t
6        sequence
7        real::p
8        real::a(4)
9        real::b(3)
10       real::c(3)
11       real::bnd
12       real::wrk1
13       real::wrk2
14     end type aos_t
15     common /aos/ aosdata
16     type(aos_t) aosdata(mimax,mjmax,mkmax)
17     !$xev end tgen src
18     !$xev tgen dst begin
19     common /pres/ p(mimax,mjmax,mkmax)
20     common /mtrx/ a(mimax,mjmax,mkmax,4), &
21           b(mimax,mjmax,mkmax,3),c(mimax,mjmax,mkmax,3)
22     common /bound/ bnd(mimax,mjmax,mkmax)
23     common /work/ wrk1(mimax,mjmax,mkmax),wrk2(mimax,mjmax,mkmax)
24     !$xev end tgen dst
25
26     !! Rule 2
27     !! Change the way of accessing the data structure
28     !$xev tgen var(i,j,k,l) exp
29     !$xev tgen var(p,a) name
30     !$xev tgen trans exp src('aosdata(i,j,k)%p') dst('p(i,j,k)')
31     !$xev tgen trans exp src('aosdata(i,j,k)%a(l)') dst('a(i,j,k,l)')
32   end program dl_himeno
```

Fig. 3 A dummy Fortran code for AoS-to-SoA convesion of the Himeno benchmark

In Fig. 3, the type definition transformation rule consists of 21 code lines. One may consider that the type definition transformation rule seems redundant because both of the original and transformed code versions are explicitly written. However, the type definition of a transformed data structure must be explicitly given by users anyway, considering the correspondence between the original and transformed types. Therefore, we believe that this is an effective and intuitive way to describe a type definition transformation rule.

In this use case, 60 code lines are in total transformed by the rules in Fig. 3. In case of converting AoS to SoA, a programmer traditionally needs to modify all of the code lines by hand. This is a tedious and error-prone task. On the other hand, in the case of using Xevtgen, user-defined code transformation can automate the task if the rule is properly defined. Moreover, the rule is reusable and/or easily customizable for another code. Accordingly, Xevtgen dummy codes can be used for accumulating and sharing expert knowledge and experiences, which can be reused by other programmers and also for other application codes.

In this article, the performance of the transformed code is compared with that of the original code using the two systems listed in Table 1. One system is the NEC SX-ACE vector computing system [8] installed at Tohoku University Cyberscience Center, and the other is a commodity PC. The former has a 10x higher memory bandwidth than the latter. Since the performance of the Himeno benchmark is usually limited by the memory bandwidth, SX-ACE can achieve a higher sustained performance than the PC if the loop is properly vectorized.

Figure 4 shows the performance impact of the data layout optimization. As shown in the figure, AoS-to-SoA conversion significantly improves the performance of each system. The performance improvement of SX-ACE is especially remarkable, because regular memory access to array elements, which are arranged in a continuous region, is more efficient than stride memory access to AoS data elements. In this use case, it is observed that the AoS-to-SoA conversion can significantly reduce memory bank conflicts and improve the memory access efficiency. This kind of performance optimization is very important to achieve high performance on a modern computing system, in which the memory access can easily become a performance bottleneck. The expressive ability of Xevtgen is high enough to express transformation rules of this important performance optimization.

It is important that different systems potentially require different data layouts. As a result, data layout optimization is likely to be system-specific and/or compiler-specific. Moreover, loop optimization might also require data layout optimization because it changes the order of accessing data elements. For example, even if SoA had a higher spacial locality of reference than AoS before optimizing a loop, SoA does not necessarily have a higher locality after the optimization. Data layout optimization often specializes an application code for a particular processor, compiler, kernel loop, and so on. Even in such a case, our Xevolver approach [1] allows the original code to remain human-friendly because system-specific and/or compiler-specific code transformations can be defined separately from the original code. Namely, system-specific and/or compiler-specific information is separated from the code.

Xevtgen enables users to achieve the separation much more easily than the original Xevolver approach. In this use case, the XSLT rules generated by Xevtgen consist of 293 lines in total. In the original Xevolver framework, users are supposed to write such XSLT rules by themselves, as reported in [9]. Writing those XSLT rules requires not only a fair amount of writing effort but also special knowledge about both XML and compilers, because users have to learn how to write XSLT rules for AST transformation. On the other hand, Xevtgen allows users to briefly express their own code transformation rules without requiring special knowledge about XML, XSLT,

Table 1 System configurations

	NEC SX-ACE	Commodity PC
Processor	SX-ACE processor	Intel Core i7 930
Peak performance	256 Gflop/s	44 Gflop/s
Memory capacity	64 GB	32 GB
Memory bandwidth	256 GB/s	25.6 GB/s
Operating system	SUPER-UX	Linux 2.6.32
Compiler	sxf90 Rev. 520	gfortran 4.4.7
Compiler options	-P auto -C hopt	-O3

and compilers. As a result, they can easily improve the performance portability of their application codes across different systems.

5 Conclusions

This article explains a code transformation rule generator, Xevtgen, for user-defined code transformations dedicated to each application code. As discussed in this article, Xevtgen allows standard users to define their own code transformations much more easily than conventional compiler-based approaches, because the users no longer need to consider code transformation rules at an AST level. They can generate code transformation rules if they know Fortran programming and several special directives.

The use case described in this article shows that Xevtgen can express code transformations required for data layout optimization, which is one of the most important code optimization techniques to exploit the performance of a modern computing

Fig. 4 Performance impacts of data layout optimization

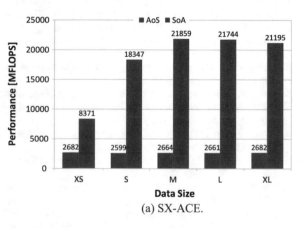

(a) SX-ACE.

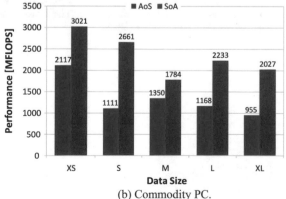

(b) Commodity PC.

system. In terms of the number of code lines, XSLT rules generated by Xevtgen are longer than a dummy Fortran code input to Xevtgen. Moreover, standard Fortran programmers would be able to describe such a dummy Fortran code if they learn how to use several special directives used by Xevtgen. In comparison with the original Xevolver approach [1] of directly writing XSLT rules by hand, the Xevtgen approach offers a much easier way of defining a practical code transformation rule.

Acknowledgements This research was partially supported by JST CREST "An Evolutionary Approach to Construction of a Software Development Environment for Massively-Parallel Heterogeneous Systems" and Grant-in-Aid for Scientific Research(B) 16H02822. The authors would like to thank all team members of the CREST project, especially Profs. Egawa, Takahashi, and Komatsu, for fruitful discussions on the design and development of the Xevolver framework.

References

1. Takizawa, H., Hirasawa, S., Hayashi, Y., Egawa, R., Kobayashi, H.: Xevolver: an XML-based code translation framework for supporting HPC application migration. In: IEEE International Conference on High Performance Computing (HiPC) (2014)
2. The Xevolver Project: JST CREST "an evolutionary approach to construction of a software development environment for massively-parallel heterogeneous systems". http://xev.arch.is.tohoku.ac.jp/
3. Quinlan, D.: ROSE: Compiler support for object-oriented frameworks. Parallel Process. Lett. **10**(02n03), 215–226 (2000)
4. Suda, R., Takizawa, H., Hirasawa, S.: Xevtgen: Fortran code transformer generator for high performance scientific codes. In: The Third International Symposium on Computing and Networking, pp. 528–534 (2015)
5. Sung, I.J., Liu, G.D., Hwu, W.M.W.: DL: a data layout transformation system for heterogeneous computing. In: Innovative Parallel Computing (InPar), pp. 1–11 (2012)
6. Kay, M.: XSLT 2.0 and XPath 2.0 Programmer's Reference (Programmer to Programmer), 4 edn. Wrox Press Ltd. (2008)
7. Himeno benchmark. http://accc.riken.jp/en/supercom/himenobmt/
8. Momose, S., Hagiwara, T., Isobe, Y., Takahara, H.: The brand-new vector supercomputer, SX-ACE. In: International Supercomputing Conference, pp. 199–214. Springer (2014)
9. Yamada, T., Hirasawa, S., Takizawa, H., Kobayashi, H.: A case study of user-defined code transformations for data layout optimizations. In: The Third International Symposium on Computing and Networking, pp. 535–541 (2015)

APES on SX-ACE

Harald Klimach, Jiaxing Qi and Sabine Roller

Abstract We report on first experiences in deploying the APES framework on the NEC SX-ACE vector system. In APES there are two solvers available, implementing different numerical schemes. Musubi is a Lattice-Boltzmann solver that can be used to simulate incompressible flows. This numerical method is attractive as it allows the explicit computation for incompressible flows with good scalability and robust treatment of highly comlex geometries. The second solver, Ateles, implements a high-order Discontinuous Galerkin method and can be used to solve hyperbolic conservation laws, including linear equations like acoustics and non-linear equations like compressible Navier-Stokes. The NEC SX-ACE vector system offers a memory bandwidth to operation ratio of 1 Byte per floating point operation, which is an interesting deployment option for many numerical schemes. Though, there are a lot of experiences with earlier systems of the SX series, the latest installment comes with new features and an overhaul of the programming environment.

1 Introduction

The APES framework [1] is a collection of applications and libraries to enable large scale numerical simulations of fluid dynamics on distributed memory systems. It is written in Fortran and utilizes some Fortran 2003 features. Though, some features from the Fortran 2003 standard are extensively used throughout the code, the development tried to stay off from various language constructs that were troublesome with various compilers over a long time. One of the notable requirements from the Fortran 2003 standard is the ISO-C-Binding, which is used to incorporate the Lua scripting language for configuration files.

H. Klimach (✉) · J. Qi · S. Roller
University of Siegen, 57076 Siegen, Germany
e-mail: harald.klimach@uni-siegen.de

J. Qi
e-mail: jiaxing.qi@uni-siegen.de

S. Roller
e-mail: sabine.roller@uni-siegen.de

© Springer International Publishing AG 2016
M.M. Resch et al. (eds.), *Sustained Simulation Performance 2016*,
DOI 10.1007/978-3-319-46735-1_4

There are mainly two fluid dynamic solvers developed within APES. One is the Lattice-Boltzmann solver Musubi [2], suitable for incompressible flows. The other is the high-order Discontinuous Galerkin solver Ateles [3] for compressible flows, but also other hyperbolic conservation laws. Both rely on a large part of shared infrastructure provided by the TreElM library [4]. Meshes are described with the help of Octree data structures, but in a sparse sense, where only elements are stored explicitly which are part of the computational domain. This sparsity results in the need for an indirection when accessing neighboring elements, like in unstructured mesh representation. The mesh format has been chosen specifically to cater large scale distributed parallel systems, besides the known topology of the Octree, a space-filling curve is used to partition the mesh. Using a space-filling curve together with the Octree enables a completely distributed partitioning, where each process can locally compute the elements it is going to work on.

Both solvers have been deployed on various large systems and have been shown to be capable of utilizing at least 100 thousand MPI processes. The APES tools have never been deployed on any earlier NEC SX vector system. Here we report some first experiences of porting this flexible framework to the NEC SX-ACE system. The NEC SX series of vector computers have a long standing tradition in high performance computing with a well-established development environment, especially for Fortran. A high memory bandwidth in relation to the floating point operation speed yields a well balanced system with 1 Byte per floating point operation. If we can exploit vector instructions in the APES solvers, we expect both solvers to benefit from the fast memory access and, thus, a high sustained performance.

Though, the APES solvers have been developed on and for cluster machines with mostly an Intel x86 architecture, vectorization was always a considered point in the development. Vectorization in the implementation is of increasing importance, as larger and larger vector instructions get into the processor architectures. The AVX2 instructions, for example, allow a single instruction to process four double precision numbers and AVX-512 extends this to eight. However, the 256 double precision numbers processed per instruction on the NEC SX processor is a completely different quality of vectorization, which requires a really strong level of vectorization in the implementation.

At least for the Lattice-Boltzmann kernel itself a good vectorization is known, due to the straight forward single loop. We therefore start out with the porting of Musubi to the NEC SX-ACE and have a look at other important parts besides the kernel. After that we move on to the little more involved Discontinuous Galerkin implementation in Ateles, where the computational effort is distributed across multiple important kernels and a vectorization is less obviously achieved.

2   Porting of Musubi

We started our porting efforts on Musubi, as this algorithm is known to perform well on the NEC SX architecture and the main kernel is straight forward to understand. The first step in porting an application is to ensure that the code can be compiled

Listing 1: No vectorization due to small intermediate arrays

```
1   do iVal=1,nVals
2     tmp(1) = a(iVal)
3     tmp(2) = b(iVal)
4
5     result = var(1) + var(2)
6   end do
```

for the target system. Luckily, the compiler environment on the NEC SX-ACE provides a Fortran 2003 compiler. This new compiler is a complete overhaul of the old compiler that only provided a very limited subset of the Fortran 2003 standard. One drawback of this new compiler is a less sophisticated optimization and vectorization, when compared to the previous more restricted compiler. Nevertheless, we found the compiler to work pretty well on Musubi and were able to tune the implementation towards a high sustained performance of about 30 % of the peak performance in the Lattice-Boltzmann kernel.

2.1 Porting of the Kernel

The Lattice-Boltzmann kernel basically is just a single loop over all lattice points. It is an explicit scheme and a double buffer is used to hold the values of the new and old time step. Therefore, there is no dependency and the expectation was a straight forward vectorization of the loop. Thanks to the NEC sxf03 compiler the initial porting without optimization was straight forward. Only very minor issues arose with some modern Fortran constructs that were easily resolved. However, the performance was disappointing and especially the vector length in vector instructions was far below the possible maximum of 256. As it turned out, the implementation of the Lattice-Boltzmann loop in Musubi made use of smaller arrays to hold temporary values. This convenience confused the compiler and had it vectorizing these arrays instead of the outer loop. To allow the compiler to put those temporary values into vector data registers, they had to be turned into individual scalars.

An illustration of the code layout where the long loop over all lattices is not vectorized is given in Listing 1. And the accordingly transformed code to allow vectorization is shown in Listing 2. For Lattice-Boltzmann the number of intermediate scalar values is usually in the range of 20 values. There explicit declaration is a little more cumbersome than employing arrays, but all in all this is a rather minor code transformation. Finally we add a hint to the compiler in Listing 2 that the outer loop iterations are indeed independent via the NODEP compiler directive. The independence is obscured by the use of indirection to address the lattices.

With the changed loop from 2, we already achieve a good vectorization and high sustained performance. For the standard collision operator, called BGK, the

Listing 2: Vectorizable loop by using scalar temporary variables

```
1   !CDIR NODEP
2   do iVal=1,nVals
3     tmp1 = a(iVal)
4     tmp2 = b(iVal)
5
6     result = tmp1 + tmp2
7   end do
```

vectorization ratio reached 99.83% with an average vector length of 256, which results in a performance of 19.87 GFLOPs on a single SX-ACE core.

For Musubi this change in code is already sufficient to achieve a satisfactory sustained performance in the Kernel on a single core. Unfortunately, for real problems we also need to consider other parts of the code, as any one of them might pose a potential bottleneck, prohibiting the execution of simulations on the machine.

2.2 Porting of the Initialization

While the kernel ran fine with the above minimal code changes, we hit a wall in the initialization for larger problems. The initialization takes care of constructing neighborhood information and setting up the indirect addressing accordingly. To achieve this a dynamic data structure is used, which allows for the addition of new elements with fast access afterwards. These are called growing arrays in Musubi and make use of amortized allocation costs by doubling the memory when the array is full.

This datatype is illustrated in Listing 3. The shown code makes use of CoCo preprocessing to define this growing array for arbitrary data types. Each array is accompanied by a counter nvals to track the actual number of entries in the possibly larger array. New entries are appended at the end of the list and the counter is increased accordingly. However, if the current size of the array is reached, the array is copied into a larger array to automatically allow for the addition of the new element. This is shown done by the shown expand routine.

For some data a dynamically growing array is not sufficient, instead it needs to be possible to search for values in the given array. To achieve this, we employ a very similar data structure, but with the addition of a ranking array to maintain a sorting of array entries and allow for binary searches. This kind of data structure has the additional complication, that for newly added values, we need to perform an insertion in the ranking array. Though the necessary copying to shift the entries is vectorizable here, the compiler needs again some hints to actually perform the vectorization.

Not too surprisingly this dynamic data structures yield only little performance on the NEC SX-ACE system, however the heavy costs getting prohibitive large for

Listing 3: Growing array to deal with dynamic data

```
1   ! Class definition
2   type grw_?tname?Array_type
3      integer :: nVals = 0
4      integer :: ContainerSize = 0
5      ?tstring?, allocatable :: Val(:)
6   end type
7
8   ! Double the array
9   subroutine expand(me, pos)
10     ...
11     me%containerSize = me%containerSize*2
12     if ( me%nVals > 0 ) then
13        allocate(swpval(me%ContainerSize))
14        swpval(:me%nVals) = me%Val(:me%nVals)
15        call move_alloc( swpval, me%Val )
16     else
17        if ( allocated(me%Val) ) deallocate( me%Val )
18        allocate( me%Val(me%containerSize) )
19     end if
20     ...
21   end subroutine expand
22
23   subroutine append(me, newval)
24     ...
25     if (me%nvals == me%containerSize) call expand(me)
26     me%nVals = me%nVals + 1
27     me%val(me%nVals) = newval
28     ...
29   end subroutine append
```

larger problems were not expected. To overcome the long running times for the initialization the following strategies have been employed:

- Avoid many small allocations: Use an initial size for the growing arrays that is in the range of expected array entries.
- Minimize the utilization of these data structures: Out of convenience the data structures where used in places where some code reorganization allowed for single allocation of fixed sized arrays instead.
- Instead of employing the data structures for arbitrary complicated derived datatype, restrict there usage to arrays of intrinsic Fortran datatypes.
- Add a NODEP compiler directive to allow the compiler the vectorization in the shifting of the ranked array.

These changes indeed cut down the initialization times to a reasonable amount, and computations of large problems became feasible. The taken steps for the Kernel and the initialization so far are crucial for all simulations. They were apparent for the most simple simulation setups, where only minimal IO had to be performed. However, for meaningful simulations it usually is necessary to load a mesh that describes a more complicated geometry. Thus, after resolving the fundamental performance issues to this point, we are now able to move on to those more involved setups.

2.3   Porting of the IO

Most simulations require the loading of meshes to describe the geometrical setup of the computation and the boundary conditions. As it turned out, the loading of meshes in Musubi was awfully slow and took in serial several minutes for a small mesh file of 32 MB. For the reading of meshes Fortran direct IO was used, which in itself so far did never pose a problem. Some further investigation revealed that the old sxf90 compiler was around 400 times faster with the same reading task as the sxf03 compiler. The explanation for this can be found in the buffering mechanism for the IO. Due to the nature of the mesh data, each read only loads 8 Bytes of data. However, the system reads 4 MB at once. The sxf90 compiler recognizes the consecutive reads and reuses the loaded 4 MB, while the new sxf03 compiler seems to not recognize it and reads the 4 for every read, resulting in the huge observed overhead. This will probably be fixed in a later release of the compiler. However, the schedule for this is not fixed yet.

To overcome this issue we now make use of MPI-IO for the reading of all data. Most IO operations already made use of MPI-IO beforehand, but for the simple reading this was not considered necessary up to now. After implementing the mesh reading also via MPI-IO, the loading time for meshes was also cut down with the sxf03 compiler.

2.4   Porting of Boundary Conditions

Related to the usage of more complicated meshes is also the treatment of boundary conditions. Indeed, after resolving the issues in the initialization and loading of the mesh, the major performance hurdle was encountered in the boundary conditions. The boundary conditions use conditionals to decide what to do in a loop over boundary lattices. As this is only badly vectorizable an alternative implementation has been put into place. To allow for a vectorization also in the boundary treatment some additional memory is introduced to maintain lists of lattices with the same boundary condition. This enables the vectorized processing of each boundary condition and finally yields an implementation that is capable of running non-trivial simulation setups with a high sustained performance on the NEC SX-ACE system.

2.5   Parallel Performance

As shown, we were able to port Musubi to the NEC SX-ACE system with relatively little effort. Missing now is the parallel performance of Musubi on the system. The NEC SX-ACE provides 4 cores per node and we employ a MPI parallelization strategy to utilize the parallelism offered by the system. For the scaling analysis we

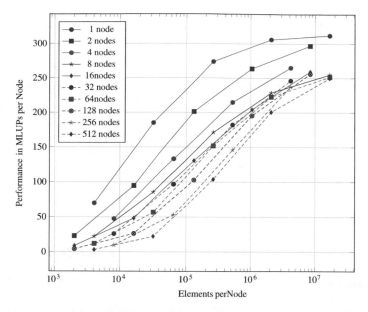

Fig. 1 Performance map of Musubi on the Tohoku NEC SX-ACE installation with up to 512 nodes (2048 MPI processes)

are using the machine of the Tohoku University in Sendai, where up to 512 nodes can be used in a parallel computation job.

To assess the parallel performance of Musubi, we use a performance map, which shows the performance per node over the problem size per node for various node counts. This graph is shown in Fig. 1. As a measure for the performance we use million lattice updates per second (MLUPs), which can also be translated into floating point operations per second. It shows a strong dependency of the performance on the problem size. For small problem sizes the performance gets diminished.

The performance map in Fig. 1 provides a full picture of the performance behavior of Musubi. We can extract the weak scaling for given problem sizes per node out of it by comparing the vertical distance between the lines for different node counts. And the strong scaling can be seen by starting on the right for a single node and moving to the left for larger and larger node counts. Thus, we recognize that weak scaling appears to rather poor, but still reasonable, while strong scaling always suffers from the strong dependency of the performance on the problem size.

Figure 2 shows the parallel efficiency when doing a weak scaling with 4 million elements per node. As can be seen the parallel efficiency drops immediately when the network is utilized (step from 1 node to 2 nodes). However, overall the drop in performance is moderate and on 512 nodes a parallel efficiency around 70 % is achieved.

A more concerning behavior is observed with respect to the performance in dependency on the problem size. Even on a single node with 4 MPI processes the

Fig. 2 Weak scaling
efficiency for 4 million
elements per node

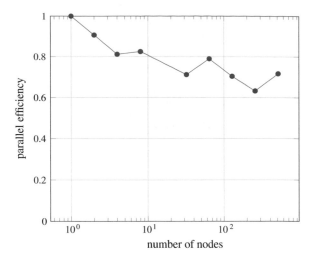

Fig. 3 Communication time
in relation to the overall
running time for increasing
problem sizes on a single
node (4 MPI processes)

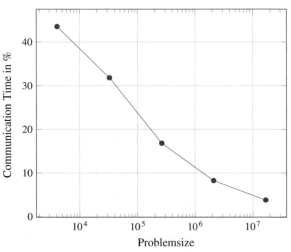

performance drops drastically for small problem sizes. This is not observed for serial
runs and we can trace this drop in performance back to the communication overhead. Figure 3 illustrates the communication overhead on a single node with 4 MPI
processes. For the smallest problem of only 4096 elements we observe a communication time of around 44 % of the overall running time. This large fraction of time
for the communication only slowly decreases for larger problems, which results in
the bad parallel performance for a wide range of problem sizes per node.

This high communication cost is probably due to a flushing of the ADB, which
becomes necessary for the MPI communication. For small problems the data from
the ADB is relatively large and the costs for purging it in the communication calls
get the dominating factor for the running time.

Unfortunately we are currently stuck with this situation as all attempts to overcome this performance bottleneck failed so far. One hope would be to avoid MPI communication by utilizing OpenMP parallelism within each node. However, the current sxf03 compiler seems to deactivate most vectorization as soon as OpenMP is activated, resulting in a poor performance in comparison to the MPI-only implementation even with the shown communication overheads. Another hope to speed the communication up was the utilization of global memory, which is available via the `MPI_alloc_mem` call of MPI. Incorporating this special memory did also not yield any benefit.

Despite the relatively poor scaling behavior due to these issues, the execution on the NEC SX-ACE compares quite well with more common large scale high performance computing systems because of the high sustained performance of nearly 30 % of the peak performance on a single node. This is illustrated in the comparison Fig. 4 by showing the sustained performance over the theoretical peak performance of utilized machine fractions in a strong scaling setup for more than 16 million elements. The comparison was done on the german systems:

- *Kabuki*: a small installation of NEC SX-ACE at the HLRS Stuttgart.
- *Hornet*: a Cray XC40 system with Intel Xeon E5-2680 v3 processors at the HLRS.
- *SuperMUC*: a Lenovo NeXtScale nx360M5 WCT system with Intel Xeon E5-2697 v3 processors located at the LRZ in Munich.
- *Juqueen*: a IBM BlueGene/Q system at the FZJ in Jülich.

As can be seen in Fig. 4, the scaling is on the NEC SX-ACE system Kabuki not as good as on the other many-core systems. However, due to the higher sustained

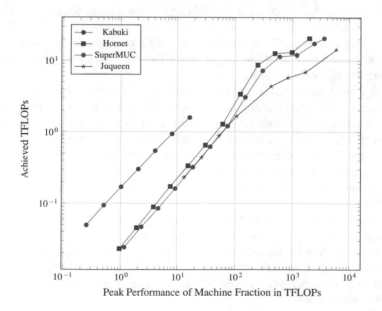

Fig. 4 Comparison of achieved performance by Musubi on different machines

performance the other systems need to utilize a much larger peak performance to obtain the same performance. Thus, even with the observed bad scaling at the moment, the NEC SX-ACE system appears to be an attractive execution option for a wide range of simulation setups. Especially, when considering the power consumption for a given simulation with Musubi, the NEC SX-ACE system shows a large advantage over the other high performance computing systems.

3   Some Notes on the Porting of Ateles

In contrast to Musubi, Ateles is not so straight forward to port to the vector system. The computational load is spread across multiple kernels and the data organization is more involved. Nevertheless, we hope to utilize the NEC SX-ACE system also with this solver and its porting is ongoing work. Though the work is not yet complete we want to share some first experiences with this Discontinuous Galerkin solver on the NEC SX-ACE.

Ateles implements a high order Discontinuous Galerkin scheme with many degrees of freedom per element. While the elements are unstructured like in Musubi, the internal organization of degrees of freedom in each element is highly regular. Our hope, therefore, is to allow a vectorized processing of the data within elements.

As a large part of the code is shared with Musubi, the initial porting without optimizations did not yield big surprises, and a first measurement could be done for a discretization of 128th order. This revealed multiple routines with significant contributions to the compute time. The three most expensive routines used respectively 32, 22 and 13 % of the compute time. For this high order, a good average vector length of 254 was achieved. However, the vector operation ratio only reached 11 % in the most expensive routine. Upon investigation, we found that a major problem in the code are again smaller loops that are put inside longer loops. Mostly these issues can be overcome by loop exchanges.

At first our attempts to remedy these problems with short loops did not yield the benefits, we would hope for. For a much smaller problem with only 16th order in the discretization scheme, the vectorization ratio only increased from 14.5 to 38 % with an average vector length of 140. A surprising discovery here was that vectorization seemed to be limited because of the size of source file. Splitting the Fortran module and using a smaller source file yielded a vectorization ratio of 99 %. This appears to be a compiler shortcoming and hopefully will be fixed in future revisions of the new compiler.

A compiler problem also seems to prohibit further vectorization of the next most important routine. This routine contains a loop, which is only partially vectorized in the new compiler, but in the old sxf90 compiler this loop gets fully vectorized. Thus, the performance porting of Ateles on the NEC SX-ACE seems to be more demanding for the vectorizing compiler and for further improvements we are looking forward to new compiler releases.

4 Summary

The porting effort of the APES solvers to the NEC SX-ACE proofed to be surprisingly smooth. For the Lattice-Boltzmann implementation in Musubi a high sustained performance of around 30 % on a single node was achieved. It has been shown that for real simulations not only the kernel needs to perform well, but also bottlenecks in the supporting infrastructure need to be overcome to allow actual simulations with complex setup. We have explained which roadblocks we encountered for the complete porting of Musubi and how they have overcome. The parallel performance has been investigated and a problem with the MPI communication was uncovered. This scalability issue remains the major shortcoming of the NEC SX-ACE for Musubi. However, despite the limited scalability, the system offers an attractive alternative to other systems due to the high sustained performance.

For the Discontinuous Galerkin solver only very first experiences could be shared, as the performance optimization got stuck early on due to a shortcoming of the current compiler version. Work on further improvements for Ateles are ongoing and we hope to be able to utilize the vectorization for the high order discretization with future compiler versions.

Acknowledgements We would like to thank Holger Berger from NEC for his ongoing kind support. This work would not have been possible without the possibility to use the NEC SX-ACE system at the Cyberscience Center, Tohoku University in Sendai and we are deeply grateful to Ryusuke Egawa to provide us with access to the system and Kazuhiko Komatsu for his support in running the jobs on the system. We also thank Uwe Küster for insightful discussions and the possibility to make use of the SX-ACE testsystem Kabuki at the HLRS in Stuttgart. Finally, we thank the Gauss-Center for supercomputing for providing the computing resources on Hornet, SuperMUC and Juqueen.

References

1. Roller, S., et al.: An adaptable simulation framework based on a linearized octree. In: Resch, M., Wang, X., Bez, W., Focht, E., Kobayashi, H., Roller, S. (eds.) High Performance Computing on Vector Systems 2011. Springer, Heidelberg (2011)
2. Hasert, M., et al.: Complex fluid simulations with the parallel tree-based lattice Boltzmann solver Musubi. J. Comp. Sci. **5**, 784–794 (2014)
3. Zudrop J., et al.: A fully distributed CFD framework for massively parallel systems. In: Cray User Group 2012. Stuttgart (2012)
4. Klimach, H., et al.: Distributed octree mesh infrastructure for flow simulations. In: Eberhardsteiner, J. (ed.) Eccomas 2012 - European Congress on Computational Methods in Applied Sciences and Engineering, e-Book Full Papers. Vienna (2012)

Dealing with Non-linear Terms in a Modal High-Order Discontinuous Galerkin Method

Nikhil Anand, Harald Klimach and Sabine Roller

Abstract The Discontinuous Galerkin (DG) method utilizes a mesh of elements with local functions like traditional continuous finite element methods, together with a flux approximation between elements like finite volume methods. This combination yields a high locality of the overall scheme, especially for high-order representations within elements. Two local operations need to be mainly considered. One is the application of the mass matrix and the other is the stiffness matrix. With an appropriate orthogonal basis as choice for the local functions both operations can be computed with minimal complexity. In this contribution we are concerned with a DG implementation that makes use of a Legendre polynomial basis with an application to non-linear equation systems. For non-linear systems a complication is introduced by the scheme by the necessity to compute the non-linear flux operation, which generally can not be done in the optimal modal basis. Instead, a pointwise evaluation of the non-linear operations is usually performed. Combining the fast evaluation of the integrals in the modal scheme with the pointwise evaluation of the non-linear terms requires a transformation between these two. Many methods have been developed for a fast transformation from Legendre modes to nodal values [1]. However, most of those methods for fast polynomial transformations are designed for extremely high polynomial degrees in the range of several hundreds. In three-dimensional DG simulations the polynomial degree in each dimension is more limited, and we are looking for methods that are fast but suitable for polynomials in the range up to a maximal degree of one hundred. We discuss some approaches to the fast transformation, especially the method proposed by Alpert and Rokhlin [2], and compare our implementation of this method to a straight forward L_2 projection. The implementation specifically addresses also the hybrid parallelism with MPI and OpenMP for the three-dimensional DG elements.

N. Anand (✉) · H. Klimach · S. Roller
University of Siegen, 57076 Siegen, Germany
e-mail: nikhil.anand@uni-siegen.de

H. Klimach
e-mail: harald.klimach@uni-siegen.de

S. Roller
e-mail: sabine.roller@uni-siegen.de

© Springer International Publishing AG 2016
M.M. Resch et al. (eds.), *Sustained Simulation Performance 2016*,
DOI 10.1007/978-3-319-46735-1_5

43

1 Introduction

High-order methods for CFD applications have gained popularity in the research community in the last decades. They have the potential to provide not only high accuracy but also efficient numerical solutions to the problems when compared to lower order methods, as the approximation error decreases exponentially with the order for smooth solutions. However, classical spectral methods suffer from limitations to simple periodic domains and a global support. Their deployment on parallel distributed systems for complex setups is usually limited in scalability.

The Discontinuous Galerkin (DG) method builds a class of schemes, that enable high-order discretizations of conservation laws. It shares many of the advantages of high-order spectral methods but overcomes its limitation of global ansatz functions by weakly coupled, element local functions. DG methods, due to this locality, provide a path towards massive parallelism with high-order on distributed systems. Within elements shared memory parallelism can be employed for the local operations, which allows the numerical scheme to match the typical hierarchy of modern computing systems.

Using an appropriate basis for the element local functions, linear operations can be efficiently computed with optimal computational complexity. For example, Legendre polynomials with their orthogonal basis and recursive definition results in a trivially invertable diagonal mass matrix and allows for a stiffness matrix that can be applied with optimal computational effort $\mathscr{O}(p^3)$ for three dimensional elements and a scheme order of p. Although an appropriate nodal basis also allows for an efficient computation of the mass matrix, the same can not be achieved for the stiffness matrix at the same time. With some restrictions the computational cost can be limited to $\mathscr{O}(p^4)$ operations in this case.

We also make use of cubical elements, which offer optimal properties for the DG scheme. The elements are organized in an Octree that enables together with a space-filling curve ordering an efficient partitioning and neighbor identification on distributed parallel systems. By restricting to hexahedral meshes in combination with orthogonal basis functions, we optimize the tensor-product nature in multiple dimensions. This enables us to use a dim-by-dim approach with minimal computational effort.

Though, linear equations can be efficiently computed with no added complexity when the appropriate basis is used, non-linear equations can generally not be treated so easily anymore. For example for nonlinear operations occurring in initial conditions, boundary conditions, source terms or non-linear fluxes, a transformation of the Legendre modes to pointwise representation needs to be performed. This forces us to look for algorithms that offer fast transformations. To allow an in-place transformation, we use the same number of points and modal coefficients. The naive evaluation of the polynomials at each of these points would result in $\mathscr{O}(p^6)$ operations in three dimensions, which clearly is not an option for high-order approximations. By employing the dim-by-dim method, the cost for this can be reduced to $\mathscr{O}(p^4)$. However, there are fast algorithms that achieve the transformation in $\mathscr{O}(p^3 log(p))$

operations theoretically. Unfortunately, many methods for fast polynomial transformations are designed for extremely high degrees and often do not exhibit their asymptotically fast behavior for low polynomial degrees like the ones used in three dimensional DG simulations. In the next paragraph, we briefly discuss some of these algorithms which we implemented in our highly parallel framework.

The Fast Polynomial Transformation (FPT) described by Alpert and Rhoklin is based on a fast approximative transformation of Legendre modes to Chebyshev polynomials [2], followed by a Fast Fourier Transformation. Another method involves the use of a Fast Multipole Method for a direct transformation of Legendre polynomials to Legendre nodes, developed by Suda [3]. This involves one algorithmic step less than the FPT. Both these method offer $\mathcal{O}(p^3 log(p))$ complexity for the transformation.

This paper is organized as follows: First, we briefly review the Discontinuous Galerkin discretization in Sect. 2. Then we highlight the choice of basis for ansatz and test function definition in Sect. 3. After that we introduce the projection algorithms and the underlying ideas in Sect. 4, followed by some strategies for hybrid parallelism using MPI and OpenMP in Sect. 5. Finally Sect. 6 presents the comparison and analysis of different projection algorithms used in this work.

2   The High Order Discontinuous Galerkin Method

In this section, we briefly introduce the semi-discrete form of the Discontinuous Galerkin Finite Element Method (DG) for compressible inviscid flows. The compressible Euler equations are the non-linear system of equations describing such flows with the conservation of mass, momentum and energy given by

$$\partial_t \mathbf{u} + \nabla \cdot \mathbf{F}(\mathbf{u}) = 0, \tag{1}$$

equipped with suitable initial and boundary conditions. Here $\mathbf{u}$ is a vector of conservative variables and the flux function $\mathbf{F}(\mathbf{u}) = (\mathbf{f}(\mathbf{u}), \mathbf{g}(\mathbf{u}))^T$ for two spatial dimensions is given by

$$
\mathbf{u} = \begin{bmatrix} \rho \\ \rho u \\ \rho v \\ \rho E \end{bmatrix}, \qquad
\mathbf{f}(\mathbf{u}) = \begin{bmatrix} \rho u \\ \rho u^2 + p \\ \rho u v \\ (\rho E + p) u \end{bmatrix}, \qquad
\mathbf{g}(\mathbf{u}) = \begin{bmatrix} \rho v \\ \rho u v \\ \rho v^2 + p \\ (\rho E + p) v \end{bmatrix},
$$

where ρ, $\mathbf{v} = (u, v)^T$, E, p denotes the density, velocity vector, specific total energy and pressure respectively. The system is closed by the quation of state assuming the fluid obeys the ideal gas law with pressure defined as $p = (\gamma - 1)\rho \left(e - \frac{1}{2}(u^2 + v^2)\right)$. where $\gamma = \frac{c_p}{c_v}$ is the ratio of specific heat capacities and e is the total internal energy per unit mass.

The Discontinuous Galerkin formulation of the above equation is obtained by multiplying it with a test function ψ and integrating it over the domain Ω. Thereafter, integration by parts is used to obtain the following weak formulation

$$\int_{\Omega} \psi \frac{\partial \mathbf{u}}{\partial t} d\Omega + \oint_{\partial \Omega} \psi \mathbf{F}(\mathbf{u}) \cdot \mathbf{n} ds - \int_{\Omega} \nabla \psi \cdot \mathbf{F}(\mathbf{u}) d\Omega = 0, \qquad \forall \psi, \qquad (2)$$

where ds denotes the surface integral. A discrete analogue of the above equation is obtained by considering a tessellation of the domain Ω into n closed, non-overlapping elements given by $T = \{\Omega_i | i = 1, 2, \ldots, n\}$, such that $\Omega = \cup_{i=1}^{n} \Omega_i$ and $\Omega_i \cap \Omega_j = \emptyset \forall i \neq j$. We define a finite element space consisting of discontinuous polynomial functions of degree $m \geq 0$ given by

$$P^m = \{f \in [L^2(\Omega)]^m\}. \qquad (3)$$

With the above definition we can write the approximate solution $\mathbf{u}_h(\mathbf{x}, t)$ within each element using a polynomial function of degree m

$$\mathbf{u}_h(\mathbf{x}, t) = \sum_{i=1}^{m} \hat{u}_i \phi_i, \qquad \psi_h(\mathbf{x}) = \sum_{i=1}^{m} \hat{v}_i \phi_i, \qquad (4)$$

where the expansion coefficients $\hat{u}_i$ and $\hat{v}_i$ denote the degrees of freedom of the approximation of solution and of test function respectively. Notice, that there is no global continuity requirement for $\mathbf{u}_h$ and ψ_h in the previous definition. Splitting the integrals in Eq. (2) into a sum of integrals over elements Ω_i, we obtain the space-discrete variational formulation

$$\sum_{i=1}^{n} \frac{\partial}{\partial t} \int_{\Omega_i} \psi_h \mathbf{u}_h d\Omega + \oint_{\partial \Omega_i} \psi_h \mathbf{F}(\mathbf{u}_h) \cdot \mathbf{n} ds - \int_{\Omega_i} \nabla \psi_h \cdot \mathbf{F}(\mathbf{u}_h) d\Omega = 0, \qquad \forall \psi_h, \qquad (5)$$

Due to element local support of the numerical representation, the flux term is not uniquely defined at the element interfaces. The flux function is, therefore, replaced by a numerical flux function $\mathbf{F}^*(\mathbf{u}_h^-, \mathbf{u}_h^+, \mathbf{n})$ where $\mathbf{u}_h^-, \mathbf{u}_h^+$ are the interior and exterior traces at the element face in the direction $\mathbf{n}$ normal to the interface.

For simplicity we can re-write the equation above in matrix vector notation and obtain

$$\frac{\partial}{\partial t} \hat{\mathbf{u}} = M^{-1} \left(S \cdot \mathbf{F}(\hat{\mathbf{u}}) - M^F \cdot \mathbf{F}(\hat{\mathbf{u}}) \right) =: rhs(\hat{\mathbf{u}}). \qquad (6)$$

where M, S denote the mass and the stiffness matrices and M^F are so called face mass lumping matrices. The above obtained ordinary differential equation (6) can be solved in time using any standard timestepping method, e.g. a Runge-Kutta method.

3 Choice of Basis Function

An important part of the DG formulation is the choice of the ansatz functions ϕ to represent the approximated solution $\mathbf{u}_h$ and test function ψ in (4). Typical choices are polynomials. This section briefly highlights the choice of the polynomial basis which allows for faster evaluation of integrals in (5). From theoritical point of view the choice of the polynomial basis is arbitrary, however, the computational effort required to evaluate the volume and surface integral term in (5) can differ based on this choice. For example, the mass matrix term $\int_{\Omega_i} \psi_h \mathbf{u}_h d\Omega$ can be cheaply computed when an orthogonal basis is choosen for both $\mathbf{u}_h$ and ψ_h. *Legendre* polynomials are orthogonal with the L_2 inner product in the interval -1 to 1. Unluckily, this orthogonality is lost for the stiffness matrix term $\int_{\Omega_i} \nabla \psi \cdot \mathbf{F}(\mathbf{u}_h) d\Omega$ since derivatives of Legendre polynomials do not simply reduce to another orthogonal basis. If the fluxes have non-linear dependence on the state (like the fluxes in the Euler equations), then this orthogonality is also lost for the surface integral term $\oint_{\partial \Omega_i} \psi \mathbf{F}(\mathbf{u}_h) \cdot \mathbf{n} ds$. For this reason, the *nodal* polynomial basis like *Lagrange* polynomials are quite common, as the coefficients can be directly identified as point values thereby allowing pointwise evaluation. The DG scheme emerging from using this type of basis are classified as *nodal DG*.

However, the evaluation of gradient is not cheap for Lagrange polynomials and when used naively the computational cost falls in the order of $\mathcal{O}(p^6)$. With some restrictions, it can be limited to $\mathcal{O}(p^4)$ operations in our case. However, for cubical domains a more efficient basis could be found which allows fast evaluation of both mass and stiffness matrix at the same time. For this we use Legendre polynomials, which is a special type of Jacobi polynomials, following a three term recursion. Being orthogonal it can cheaply evaluate the mass matrix and the recursion definition helps to assemble the stiffness matrix in $\mathcal{O}(p^3)$. The DG scheme based on these kind of basis functions for approximation are known as *modal DG*. Apart from this, modal DG has also other cheaper and efficient means when it comes to dealing with aliasing errors, filtering and stabilization techniques or fast projection of solution onto faces etc. We skip further details as it is not in the scope of this paper.

4 The Projection Algorithms

The Legendre polynomial series does not have a fast transform associated with it like a Chebyshev expansion. Therefore, a $p + 1$-th order Legendre expansion would normally require $\mathcal{O}(p^2)$ operations to evaluate point values at p nodes. With multiple dimensions this high computational cost gets significant even for low orders, as the polynomials in all directions need to be considered, resulting in $\mathcal{O}(p^{2d})$ operations for d dimensions. With a tensor product formulation on cubical elements, a dim-by-dim algorithm can be deployed and the number of iterations reduced to $\mathcal{O}(d \cdot p^{d-1} \cdot p^2) = \mathcal{O}(p^{d+1})$. Without loss of generality, but for readability, we restrict ourselves to a single dimension here. Next, we give a formal definition of the problem of polynomial projection and then go on describing the algorithms we use along with relevant implementation details in the subsequent subsections.

Given is a function $f : [-1, 1] \to \mathbb{R}$ expressed by an n term finite Legendre expansion of the form

$$f(t) = \sum_{i=0}^{n-1} \hat{u}_i \cdot L_i(t). \tag{7}$$

We want to convert this expansion to n nodal point values of f evaluated at the points $t_1, t_2, \ldots, t_n$. Similarly, for the inverse operation, given tabulated values of function $f : [-1, 1] \to \mathbb{R}$ at n nodes $t_1, t_2, \ldots, t_n$ we want to evaluate the coefficients $\hat{u}_0, \hat{u}_1, \ldots, \hat{u}_{n-1}$ such that

$$f(t_j) = \sum_{i=0}^{n-1} \hat{u}_i \cdot L_i(t_j) \tag{8}$$

holds. These transformations in the above noted general formulation require $\mathcal{O}(n^2)$ operations. For a polynomial representation in 3D the number of operations required for this transformation would be proportional to n^6. Thus, a naive implementation of the general transformation is prohibitively expensive even for moderately large polynomial degrees n. This creates the need to look for an efficient way to transform modal expansions to point values and back in order to retain high performance for higher order.

Alpert and Rokhlin presented a fast and stable algorithm for fast transformations of Legendre polynomials [2]. Various alternatives and extensions were proposed since then e.g [1, 3]. A detailed analysis of fast polynomial transformations can be found in [4]. Figure 1 shows some existing fast algorithms that can deliver fast transformation from modal coefficients to point values and back. The solid line in the figure higlights the transformation algorithm we implemented tailored to fit our highly parallel framework. In the following subsections we briefly discuss the projection algorithms and its implementation details. Then we compare the algorithms and analyse them.

4.1 Direct L_2 Projection

The direct, but expensive transformation between the nodal and modal basis is the L_2 projection, which refer to as L2P in this paper. Mathematically, the L_2 projection f_h, for Legendre expansion f of the form (7) projected onto any arbitrary function space $\theta \in L_2(\Omega)$ is given by

$$\langle f_h - f, v \rangle_{L_2(\Omega)} = 0 \qquad\qquad \forall v \in \theta, \tag{9}$$

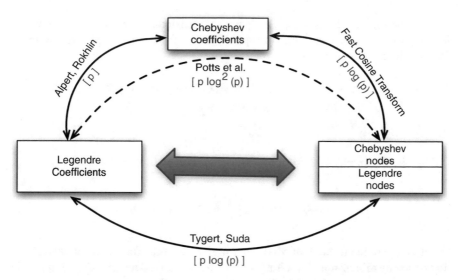

Fig. 1 This figure shows some of the available fast algorithms for converting Legengre coefficients to point values along with their expected computational complexity for transformation in single dimension marked in *red*. *Solid arrows* highlight the algorithms we used in this work

or equivalently

$$\langle f_h, v \rangle_{L_2(\Omega)} = \langle f, v \rangle_{L_2(\Omega)} \qquad\qquad \forall v \in \theta. \qquad (10)$$

Then, for each element the discrete counterpart of this system can be stated as

$$Mx = b, \qquad (11)$$

where components of matrix $M_{i,j} = \langle v_j, v_i \rangle$, $\{v_i\}$ being basis of space θ, x is the vector containing degrees of freedom of $f_h = \sum_i x_i v_i$ and components of b_i is given by $b_i = \langle f, v \rangle$.

By choosing the Lagrange polynomials as target space, this approach be be used to transform the represenation from Legendre modes to Legendre nodes and back. For an accurate mapping and to allow in-place transformation, we use as many points as modal coefficients. As can be easily seen from the matrix formulation the costs of this transformation grows quadratically with the number of degrees of freedoms. Furthermore, the costs increase also with the dimensionality of the polynomial. For example, a three dimensional p-th order element has p^3 coefficients for each variable. Evaluating it at all the p^3 nodes would take p^6 operations, which would quickly get prohibitive expensive for higher orders. However, tensor product basis functions and cubical elements help us to reduce the problem formulation to multiple one dimensional operations, which in general is not possible for non tensor-product elements like tetrahedral elements.

Considering directions $k \in \{1, 2, 3\}$ for 3 dimensions and order p representation in an element Ω_i we can define the spatial ansatz function as a product of ansatz function in one dimension by

$$\phi_l(\mathbf{x}) = \phi_l(x_1, x_2, x_3) = \phi_{l_1}(x_1) \cdot \phi_{l_2}(x_2) \cdot \phi_{l_3}(x_3)$$
$$l = 1 + l_1 + l_2 \cdot p + l_3 \cdot p^2$$
$$1 \leq l_1, l_2, l_3 \leq p^3$$

Using this, the terms of the form $\langle \phi_a(\mathbf{x}), \phi_b(\mathbf{x}) \rangle$ can be broken down into

$$\langle \phi_a(\mathbf{x}), \phi_b(\mathbf{x}) \rangle = \prod_{k=1}^{3} \langle \phi_{a_k}(x_k), \phi_{b_k}(x_k) \rangle \,,$$

From computational point of view, this makes the transformation applicable for dimension by dimension. Exploiting this we are able to reduce the complexity of the L_2 projection down to $\mathcal{O}(p^4)$ in three dimension.

4.2 Fast Polynomial Transformation

The FPT algorithm by Alpert and Rokhlin [2] is based on the idea to exploit the already known fast transformation for Chebyshev polynomials in the form of the fast cosine transform. As we are looking for the transformation of a Legendre expansion (7), the missing component is a fast transformation between Legendre and Chebyshev coefficients. In this section we describe the basic concept and implementation for this fast polynomial transformation, which we refer to as FPT.

Assuming, a function can be described by a finite Legendre expansion as given in (7) it then can also be described by a finite Chebyshev expansion of the form

$$f(x) = \sum_{i=0}^{n-1} \hat{u}_i^c \cdot T_i(x) \tag{12}$$

where $T_i(x)$ is the i-th Chebyshev polynomial. The coefficients $\hat{u}_i^l$ and $\hat{u}_i^c$ are then related by the equation

$$\hat{u}^c = M \cdot \hat{u}^l \tag{13}$$

where, $\hat{u}^c = (\hat{u}_0^c, \hat{u}_1^c, \ldots, \hat{u}_{n-1}^c)$ and $\hat{u}^l = (\hat{u}_0^l, \hat{u}_1^l, \ldots, \hat{u}_{n-1}^l)$. Also, conversely, if f is a function defined as the Chebyshev expansion in (12) then it can be expressed as the Legendre expansion of the form (7), with the coefficients $\hat{u}^l$ given by

$$\hat{u}^l = L \cdot \hat{u}^l \tag{14}$$

Alpert and Rokhlin showed that the entries of matrices M and L can be expressed by meromorphic functions of the matrix indices and have the following structure:

$$
M_{i,j} = \begin{cases} \frac{1}{\pi} \Lambda(j/2) & \text{if } 0 = i \leq j < g+1 \text{ and } j \text{ is even} \\ \frac{2}{\pi} \Lambda\left(\frac{j-i}{2}\right) \Lambda\left(\frac{j+i}{2}\right) & \text{if } 0 < i \leq j < g+1 \text{ and } j+i \text{ is even} \\ 0 & \text{otherwise} \end{cases} \quad (15a)
$$

$$
L_{i,j} = \begin{cases} 1 & \text{if } i = j = 0 \\ \sqrt{\pi}/(2\Lambda(i)) & \text{if } 0 < i = j < g+1 \\ \frac{-j(i+1/2)}{(j+i+1)(j-i)} \Lambda\left(\frac{j-i-2}{2}\right) \Lambda\left(\frac{j+i-1}{2}\right) & \text{if } 0 \leq i < j < g+1 \text{ and } i+j \text{ is even} \\ 0 & \text{otherwise} \end{cases}
$$

$$(15b)$$

The function $\Lambda : \mathbf{C} \to \mathbf{C}$ is defined by $\Lambda(z) = \Gamma(z+1/2)/\Gamma(z+1)$. Therefore, blocks in the matrices can be approximated cheaply by e.g. a Taylor series expansion using only a few coefficients. The farther away the blocks from the diagonal, the less accurate their approximation needs to be. Thus, approximation costs can drop with distances to the diagonal. For further details and analysis we refer to [2] The matrices M and L are subdivided into entries close to the diagonal, triangle submatrices and blocks. Diagonals and triangles are evaluated directly, while for the blocks only an approximation is used. The block-size grows with the distance from the diagonal. A sketch of this decomposition of the matrices is given in Fig. 2.

Due to the fact that the number of rows scales proportional to p and the number of blocks scale as $\mathcal{O}(log(p))$, the computational complexity of the algorithm is $\mathcal{O}(p\, log(p))$ as a constant effort for all blocks is used no matter their size.

Also, this algorithm is more stable than the direct L_2 projection as the scheme does not give rounding errors when the points are located close to the boundary of the reference element.

4.3 Spherical Harmonic Transform Using Fast Multipole Method

Spherical harmonics are set of spatial functions forming orthogonal system defined on the surface of a sphere. Several transformation exist for performing spherical harmonic transforms [5–7]. Suda proposes a fast transformation algorithm for the same using the Fast Multipole Method (FMM) [7, 8]. In this section we describe the basic idea of algorithm proposed by Suda.

Spherical harmonic can be expressed as product of an associated Legengre function and a trigonometric function. This way, the spherical harmonic transform breaks down into successive evaluation of an associated Legendre transform and an inverse

Fig. 2 Subdivision of
matrices M and L. Entries
close to the diagonal are
directly applied while the
blocks further away are
approximated using a Taylor
series expansion. The
triangles are also computed
directly, just like the
diagonals

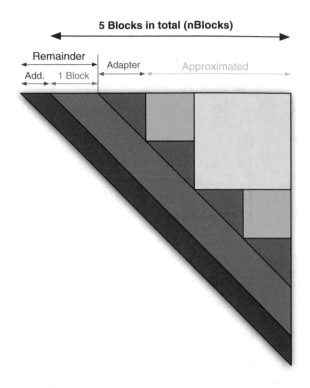

fourier transform. Fast algorithm for the latter already exists, so it is sufficient to
accelerate the associated Legendre transform for fast spherical harmonic transform.
Suda [7] uses polynomial interpolation to evaluate associated Legendre transform
and accelerates the polynomial interpolation using Fast Multipole Method (FMM).
For detailed algorithm we refer to [7].

A spherical Harmonic function $Y_j^k(\lambda, \mu)$ can be represented as a product of an
associated Legendre function and a trigonometric function as

$$Y_j^k(\lambda, \mu) = L_j^k(\mu)e^{ik\lambda} \tag{16}$$

where μ and λ are spherical angular coordinates. L_j^k is the associated Legendre
function of degree j and order k. The evaluation of the spherical harmonic expression

$$g(\lambda, \mu) = \sum_{k=1}^{p+1} \sum_{j=k}^{p+1} g_j^k Y_j^k(\lambda, \mu) \tag{17}$$

can also be split into an associated Legendre function transform and a Fourier transform as

$$g^k(\mu) = \sum_{j=k}^{p+1} g_j^k L_j^k(\mu), \tag{18}$$

$$g(\lambda, \mu) = \sum_{k=1}^{p+1} g^k(\mu)e^{ik\lambda} \tag{19}$$

The inverse spherical harmonic transform involves the computation of $g(\lambda_r, \mu_s)$ from the input coefficients g_j^k, while the spherical harmonic transform is the computation of g_j^k from sampled values from $g(\lambda_r, \mu_s)$. For the so-called alias free condition, the indices are restricted in the following way

$$\{r \in (1, R), s \in (1, S); R \geq 3p + 1, S \geq \frac{3p + 1}{2}\}. \tag{20}$$

Thus the inverse transform consists of associated Legendre function transform (18) and Fourier transform (19). The Fast Fourier Transform already enables (19) to be computed optimally in $\mathcal{O}(p \, log(p))$. Thus accelerating the associated Legendre function transform is sufficient enough to reduce the complexity of the whole transformation. Suda proposes an algorithm based on this idea, and the computational complexity of the whole transformation is $\mathcal{O}(p \, log(p))$ for 1D. There is also a set of routines publicly available as a C library to perform these transformations [9]. We integrated the FXTPACK routines into our program and use it to perform transformations.

5 Hybrid Parallelization of the Projection Algorithms

As we discussed earlier, a high locality coming from loosely coupled elements in DG is key to good scalability on distributed systems. Also, the workload per element gets high with increasing order. For example, for p-th order scheme there are p^3 unknows per variable per element. However, this workload can not be distributed among different processes efficiently, as access pattern within an element is quite random and tightly coupled. However, with shared memory parallelization of operations within each element the scalability can be increased, especially for high order discretizations. So, the elements can be distributed among the processes and within elements shared memory parallelism can be deployed. Using this technique, its possible to scale down to one element per node. We used OpenMP to parallelise the L2P projection algorithm. We present and discuss the results for hybrid parallelization in Sect. 6.

6 Comparison of Different Algorithms

In this section we discuss the comparison of different transformation algorithms, highlighting the expected and achieved performances. Then we go on presenting some results on the hybrid parallelization as discussed in Sect. 5. Finally, we briefly analyse the performance behavior with respect to vectorization and OpenMP.

As a testcase we use the compressible Euler equation in 3D to simulate the flow of a fluid in a simple cubic domain with periodic boundaries. For time stepping, we use 4-step explicit Runge-Kutta (RK4) method. Because of the non-linear flux term the computation requires the conversion from modal to nodal coefficients and back in the otherwise modal scheme. With explicit RK4 time stepping, it needs to do this transformation 4 times for each single time step iteration and for each conservative variable (a total of 5 conservative variables).

To measure the performance, we consider the whole computation loop. There are, of course, several other operations apart from polynomial transformations contributing to the overall performance. However, for sufficiently high orders the transformations are the most significant factor. And also for the overall behavior, which is in the end the relevant measure, the impact of the transformation performance can be observed. With all other simulation parameters remaining the same, we believe the overall performance is a valid indication for a comparison between the different transformation algorithms in the actual simulation setup. We use the measure of thousand degree of freedom updates per second (KDUPS) for the performance assessment. A degree of freedom update refers to the time taken to update a single degree of freedom from one timestep to the next. Larger KDUPS imply better performance and vice versa. Also notice that the performance attained includes effects of the implementation and the computing hardware, such as caching or vectorization.

In Fig. 3 we plot the performance of the different transformation algorithms. First in Fig. 3a we measure the performance on our small development platform, which uses intel Xeon X5650 (Westmere) processors. The performance index KDUPS is plotted against the increasing order on the horizontal axis. The overall problem size is kept nearly constant around 80 million degrees of freedom per variable. Thus, with an increasing order the mesh resolution gets coarser to maintain the overall problem size (or total number of degrees of freedom). The peak in the low order range shows a caching effect, where a single elements can be kept completely in the cache. There, the computation is faster as it benefits of the data locality for all operations inside the element. For higher orders this effect gets lost as data needs to be fetched from memory even for element local operations. The performance flattens out. On this machine we observe that that after the 8th order the performance of FPT is better when compared to others, even though the asymptotic fast regime seems to be achieved only for very high orders. At an order of 256 we observe a small dip in the performance of the FPT, but apart from that the FPT always appears to be the fastest option. Running exactly the same setup on an Intel Xeon E5-2680v3 (Haswell) processor, we observe a different behavior. While the performance for all transformation methods improves due to the faster processor, we can also observe a speed-up for L2P, which becomes

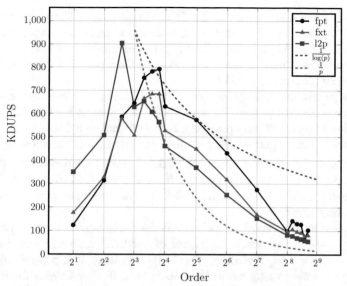

(a) Comparison of projetion algorithms on intel Xeon X5650 (Westmere) processor

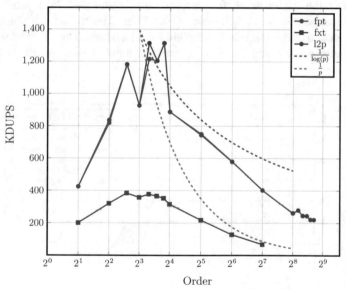

(b) Comparison of projetion algorithms on Intel Xeon E5-2680v3 (Haswell) processor

Fig. 3 Performance measure of the transformation algorithms. L2P denotes the direct L2 projection, FPT is the fast polynomial transformation and FXT is the spherical harmonic transform of the FXTPACK library

comparable to the performance of the FPT. This benefit is likely due to the well vectorizable parts of the L2P algorithm, which is of increased importance on the newer processor. At the same time, the FXT implementation can not profit from the improved performance on the newer processor, which is probably due to the irregular memory access patterns in the FMM.

We observe the direct L_2 projection with the dim-by-dim optimization yields a peformance equivalent to the FPT up to scheme orders as high as 400. Though, the FPT should asymptotically provide a computational complexity of $p^3 \, log p$ and the L2P of p^4. These operation estimations correspond to a line following $\frac{1}{log p}$ and $\frac{1}{p}$ respectively. The expected asymptotic behaviors are included in the figures for reference. As we can see, it is hard for the fast algorithms to compete with the simple direct transformation, which just inflicts a matrix-vector product that can be computed very efficiently.

As we mentioned in Sect. 5, the DG scheme can exploit shared memory parallelism for higher orders as the number of degrees of freedom within an element increases and with them the data parallelism. Due to its simplicity, it is possible to trivially parallelize the matrix vector multiplications in the L2P. Fig. 4 shows the intra-node performance with OpenMP and MPI.

Plotted is the performance for various combinations of MPI processes and OpenMP thread counts on a single node of Hazel Hen. The problem size (approx. 800 million degrees of freedom) is kept constant for all the runs. Thus with an increasing spatial scheme order along the X-axis, the total number of elements in the mesh

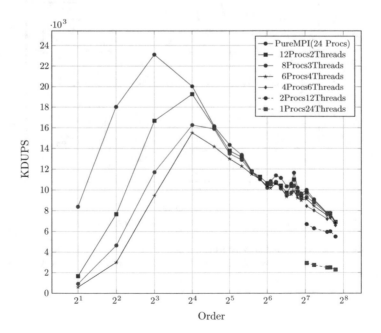

Fig. 4 OpenMP performance of the L2P for various process-thread counts

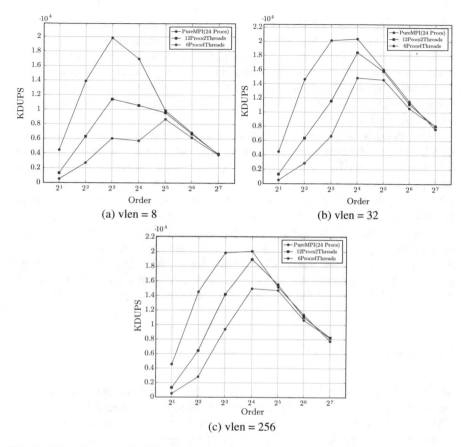

Fig. 5 L2P performance for different proc-thread combination with varying vector length from **a**, **b**, **c**

decreases. Also, the number of elements used is always a multiple of 24, such that it can always be evenly distributed among the up to 24 MPI processes on the single node. This ensures there is no load imbalance due to different number of elements on each MPI ranks and thereby the pure MPI performance does not get distorted. The run with the highest order in the graph makes use of 24 elements. We would expect the performance of hybrid runs to be close to pure MPI runs so that it would enable us to use OpenMP without loosing performance. In Fig. 4 we see the performance of pure MPI is clearly better for scheme orders up to 16. This is expected as for 16th order (or 4096 degrees of freedeom per variable and element), the computational load within a single element is relatively low and insufficient to break even with the overheads introduced by the OpenMP parallelization. However, we can see that the performance of hybrid runs closes in to MPI-only computations for increasing scheme orders. We observe using 4 threads gets us quite close to the performance obtained using 24 MPI processes for high orders. At the same time, it doesn't pay off to use large number of threads (>6) as the performance deteoriates as soon as threads span multiple NUMA domains. Also, its worth mentioning that hybrid parallelization allows us to fit larger problem per element within a node still attaining optimal performance. For example, when hit the memory bound with a certain order on a node, with shared memory parallelisn we can reduce the number of elements per node to half and use 2 OpenMP threads instead and use even higher orders, while still utilizing all cores.

Many modern hardwares offer possibilities of speeding up the computation using data level parallelization with vectorization. We also exploit this feature inside our L2P implementation by vectorizing our loop operations. We perform matrix vector multiplication using chunks or vectors of specific length. This vector length can be set up during the compilation and helps us boost the performance on vector machines. OpenMP parallelism is implemented on this vector chunks. Thus, it needs to be tuned to obtain the optimum performance on a given system. Figure 3 shows the OpenMP performance for 3 different vector lengths. When the vector size is too small (e.g. Fig. 5a), we see that the OpenMP overheads are too high and, therefore, a larger difference in performance. As we increase the vector length, the OpenMP threads get more work and we see the improvement in the performance of hybrid runs.

7 Conclusion

In this work we presented some of the fast algorithms available as an option to efficiently transform between modal and nodal spaces specially needed when dealing with non-linear terms in modal high order Discontinuous Galerkin methods. We discussed the implementation aspects and the performance comparison of the algorithms we implemented in our code. Then we also talked about hybrid parallelizing the DG method and presented some performance plots highlighting efficient implementation. However, we did not find the performance of the fast algorithms convincing especially for lower orders. They mostly start to pay off for orders which are not feasible for 3D simulations. We found the L2P algorithm quite handy and a decent option since it is easy to optimise because of its simple structure. We still look out for some fast algorithms paying off for order less then hundred. We are further working on implementing shared memory parallelism of our FPT implementation and make it suit our framework and exploit dimension by dimension approach.

Acknowledgements The performance measurement were performed on the Hornet supercomputer at the High Performance Computing Center Stuttgart (HLRS). The authors wish to thank for the computing time and the technical support.

References

1. Hale, N., Townsend, A.: A fast, simple, and stable Chebyshev-Legendre transform using an asymptotic formula. SIAM J. Sci. Comput. **36**(1), A148–A167 (2014)
2. Alpert, B.K., Rokhlin, V.: A fast algorithm for the evaluation of legendre expansions. SIAM J. Sci. Stat. Comput. **12**(1), 158–179 (1991). doi:10.1137/0912009
3. Potts, D., Steidl, G., Tasche, M.: Fast algorithms for discrete polynomial transforms. Math. Comp. **67**(224), 1577–1590 (1998). doi:10.1090/S0025-5718-98-00975-2
4. Iserles, A.: A fast and simple algorithm for the computation of legendre coefficients. Numerische Mathematik **117**(3), 529–553 (2011). doi:10.1007/s00211-010-0352-1
5. Mohlenkamp, M.J.: A fast transform for spherical harmonics. J. Fourier Anal. Appl. **5**(2), 159–184 (1999). doi:10.1007/BF01261607
6. Schaeffer, N.: Efficient spherical harmonic transforms aimed at pseudospectral numerical simulations. Geochem. Geophys. Geosyst. **14**, 751–758 (2013). doi:10.1002/ggge.20071
7. Suda, R., Takami, M.: A fast spherical harmonics transform algorithm **71**(238), 703–715 (2002)
8. Suda, R., Kuriyama, S.: Another preprocessing algorithm for generalized one-dimensional fast multipole method. J. Comput. Phys. **195**(2), 790–803 (2004). doi:10.1016/j.jcp.2003.10.018
9. Suda, R.: Fxtpack. http://sudalab.is.s.u-tokyo.ac.jp/~reiji/fxtpack.html

Efficient Coupling of Fluid and Acoustic Interaction on Massive Parallel Systems

Verena Krupp, Kannan Masilamani, Harald Klimach
and Sabine Roller

Abstract We present and compare two coupling approaches for direct aeroacoustic simulations. Direct aeroacoustic simulations pose a multi-scale problem, as the generation of sound in a flow field occurs at small spatial scales with high energy, while its propagation in the farfield has to be observed on a large spatial scale with only low energy. The challenge of different scales can be addressed by employing different numerical schemes in the individual spatial areas with an interaction between them on the surfaces. Two implementation strategies of this coupling approach are presented. The first coupling strategy employs a library that allows a wide range of different applications to be coupled with minimal changes to the individual solvers. Hence, this is a very flexible approach but limited access to information and therefore cope with loss of potential performance. Further this strategy involves the handling of multiple executables on today supercomputer. This multi-solver approach requires data interpolation at the coupling interface which introduce another numerical error. In contrast, the second approach is fully integrated within one numerical framework. Thereby the solvers are invoked as a library by the coupling application and only one single applications must be handled. Tethering high order solvers, fully access to the data implies that no additional data interpolation is required which promise better numerical results. This tight integration allows for the exploitation of knowledge about internal data structures and therefore yield performance benefits accompany with less flexibility. Both strategies will be compared with respect to numerical error due to data interpolation at the coupling interface as well as scalability and performance on modern supercomputer.

V. Krupp (✉) · K. Masilamani · H. Klimach · S. Roller
University of Siegen, Hölderlinstr. 3, Siegen, Germany
e-mail: verena.krupp@uni-siegen.de

K. Masilamani
e-mail: kannan.masilamani@uni-siegen.de

H. Klimach
e-mail: harald.klimach@uni-siegen.de

S. Roller
e-mail: sabine.roller@uni-siegen.de

© Springer International Publishing AG 2016
M.M. Resch et al. (eds.), *Sustained Simulation Performance 2016*,
DOI 10.1007/978-3-319-46735-1_6

1 Introduction

Increasing computational resources allow the simulation of a new range of multi-physics and multi-scale problems that were unfeasible with former compute resources. Such simulations have the potential to provide more insight into applications from various fields, as, for example, the sound design of aircrafts or wind turbines. With an increased awareness for noise pollution such consideration get more and more important in the design process of industrial applications.

In this work, we focus on the coupling of fluid flows and acoustic sound propagation. The main challenge of this coupled application is that both phenomena include different length and energy scales. The multi-scale nature of the fluid-acoustic interactions is best described in the example of a wind turbine: Noise is generated by the vortices of the rotating geometry at a length scale in the order of centimeters. The whole turbine size is in the scale of meters, while the noise emission is of relevance in a distance of hundreds of meters up to a few kilometers from the sound source. Simulating the entire domain while resolving the smallest turbulent scales and resolving the boundary layer adequately would require approximately 10^{18} degrees of freedom, which is out of reach even with the larges computing facilities in the forseeable future. For fluid-acoustic interactions the phenomena can often be clearly separated in the different areas of the domain. Different sets of equations and different discretization resolutions and schemes can be used for each part individually. The fluid-acoustic coupling interface is rather large and, therefore, needs to be efficient and fully parallelized.

We describe a partitioned coupling approach, i.e., we split the physical space into smaller domains, each covering a so-called single-physics subdomain. These subdomains can be solved with numerical methods and resolutions tailored to the local physical requirements. This allows for the re-use of existing scalable software based on decades of experience in each single-physics discipline, thus enabling acceptable software development times along with efficiency and performance optimization. The interaction between the domains is realized by exchanging data at the boundary. By the adaptation of numerical approximations in the individual domains, the computation of complete interactions between fluid mechanics and acoustic wave propagation becomes feasible.

In this paper, we investigate two different partitioned coupling approaches. One makes use of individual solvers that run as independent executables and use a coupling library to exchange data. The other approach uses a more integrated approach, where a single application is used and the individual solvers are incorporated as libraries. This tight integration on the basis of a common framework allows for the exploitation of knowledge about internal data structures and therefore potentially a faster coupling mechanism. However, this comes at the cost of reduced flexibility. The presented work focuses on establishing both approaches within the simulation framework *APES* and compares numerical as well as performance results. First, we briefly recapitulate the governing equations for fluid mechanics and acoustic wave propagation in Sect. 2 followed by Sect. 3 describing the methodology of the flow and acoustic solver *Ateles*. Section 3.2 describes the partitioned coupling approach in general including the multi-solver approach using the open-source coupling library *preCICE*, and the integrated coupling approach *APESmate* within the numerical framework *APES*. Finally, Sect. 4 presents the results of numerical simulations of two academic testcases as well as performance results for both approaches.

2 Governing Equations

Acoustic phenomena are based on the same principals as fluid motion. However, while for general fluid motion nonlinear equations have to be considered, acoustic phenomena can be represented in linearized equations, as only small perturbations need to be considered. The linearization reduces the numerical effort drastically and, therefore, is a necessity for large computational domains as required for the computation of acoustic far fields.

2.1 Fluid Equations

Frictionless flow is governed by the compressible Euler equations based on the conservation of mass, momentum and energy. We use the superscript f to indicate variables in the flow field. The conservation of mass can be written as

$$\frac{\partial \rho^f}{\partial t} + \nabla \cdot (\rho \mathbf{v})^f = 0, \tag{1}$$

the conservation of momentum is given by

$$\frac{\partial (\rho \mathbf{v})^f}{\partial t} + \nabla \cdot \left((\rho \mathbf{v})^f \mathbf{v}^f \right) + \nabla p^f = 0, \tag{2}$$

and the conservation of energy yields

$$\frac{\partial}{\partial t} \left(\rho^f \left(e^f + \frac{1}{2} \mathbf{v}^f \cdot \mathbf{v}^f \right) \right) + \nabla \cdot \left((\rho \mathbf{v})^f \left(e^f + \frac{1}{2} \mathbf{v}^f \cdot \mathbf{v}^f + \frac{p^f}{\rho^f} \right) \right) = 0. \tag{3}$$

The velocity field is denoted by $\mathbf{v}^f$, pressure is denoted with p^f, and the density is given as ρ^f. The internal energy of the flow is e^f. The Euler equations are derived from the Navier-Stokes equations by neglecting viscous effects, heat flow and external forces. We only consider ideal gases here to close the system:

$$p^f = \rho^f R T = (\gamma - 1) \left(e^f - \frac{\rho^f \mathbf{v}^f \cdot \mathbf{v}^f}{2} \right)$$

which yields a relation between pressure p and energy e, where R is the ideal gas constant, T is the temperature and γ is the isentropic coefficient.

2.2 Acoustic Equations

Acoustic phenomena are also fluid motion and are, therefore, governed by the Euler equations (1)–(3). As there are only small changes in the flow, they can be linearized around a constant background flow. The constant background flow is denoted by the subscript 0 and the perturbation is denoted with the superscript a. In the following, we will treat only the primitive variables density ρ, velocity $\mathbf{v}$ and pressure p in the acoustic domain. The linearized Euler equations are given by the linearized equation of mass conservation

$$\frac{\partial \rho^a}{\partial t} + \nabla \cdot \left(\mathbf{v}_0 \rho^a + \rho_0 \mathbf{v}^a\right) = 0, \tag{4}$$

the linearized momentum equation

$$\frac{\partial \mathbf{v}^a}{\partial t} + \nabla \cdot \left(\mathbf{v}_0 \mathbf{v}^a + \frac{1}{\rho_0} p^a\right) = 0 \tag{5}$$

and linearized energy equation

$$\frac{\partial p^a}{\partial t} + \nabla \cdot \left(\mathbf{v}_0 p^a + \gamma\, p_0\, \mathbf{v}^a\right) = 0. \tag{6}$$

Since the Euler equations require conservative variables for the coupling, the general transformation between primitive variables $\rho^f, \mathbf{v}^f, p^f$ and conservative variables $\rho^f, \rho^f \mathbf{v}^f, e^f$ is required

$$\rho^f = \rho^f, \quad \mathbf{v}^f = \frac{(\rho \mathbf{v})^f}{\rho^f}, \quad p^f = (\gamma - 1)[\rho^f e^f - \frac{1}{2\rho^f}((\rho \mathbf{v})^f)^2],$$

as well as vice versa

$$\rho^f = \rho^f, \quad (\mathbf{v}\rho)^f = \mathbf{v}^f \rho^f, \quad e^f = \frac{1}{(\gamma - 1)} \frac{p^f}{\rho^f} + \frac{1}{2}(\mathbf{v}^f)^2.$$

To compute the linearized variables in the acoustic domain, simple subtraction of the background state is sufficient to obtain the perturbations:

$$\rho^a = \rho^f - \rho_0, \quad \mathbf{v}^a = \mathbf{v}^f - \mathbf{v}_0, \quad p^a = p^f - p_0,$$

where the background flow is defined by the user.

3 Methodology

In this section, we describe the methodology of the established simulation approach. First we present the flow and acoustic solver *Ateles*. Then, we describe the partitioned coupling approach for fluid-acoustic interaction in detail including the two implementation strategies: the multi-solver approach which uses the open-source coupling library *preCICE* and the integrated approach *APESmate* which is incorporated within the framework *APES*.

3.1 *High-Order Solver* Ateles

For the flow as well as the acoustic domain, we use the high-order solver *Ateles* which is included in the end-to-end parallel framework *APES* [6, 10]. The *APES* framework is designed to take advantage of the massively parallel systems available in supercomputing today. Therefore, it provides additional tools for pre- and post-processing on the basis of a common mesh library.[1] The *TreElM* library [4] relies on an octree representation of the mesh and provides a distributed neighborhood search within that mesh. Using a space-filling curve for the domain decomposition of the octree mesh gives hierarchically structured data and maintains locality. This locality can be perfectly exploited by the high-order Discontinuous Galerkin solver *Ateles*.

Ateles is capable to solve various equation systems such as compressible flow, linear wave propagation and electro-dynamics, which are solved with an explicit Runge Kutta method in time and a modal discontinuous Galerkin method (DG) with arbitrary order in space [3]. The Discontinuous Galerkin method is based on a polynomial representation within an element and flux calculation between elements over their faces. Hence, there is a strong coupling of data within each element and only a loosely coupling between elements via element surfaces. The choice of the polynomial degree controls the spatial discretization order. By choosing a high degree for the polynomial function a high-order method is constructed. Exploiting modal basis functions has computational reasons, e.g. that the numerical flux can be directly evaluated in modal space, using cubical elements without any extra transformation to a reference element [9].

A higher order scheme has several advantages. First, it yields low numerical dissipation and dispersion errors, which is advantageous for approximating the wave propagation over long distances in the acoustic far field. Secondly, a higher order scheme shows high convergence rates in case of smooth solutions. Hence, a high order approximation provides a high accuracy with only few degrees of freedom. For nonlinear systems high-order schemes imply an increased computational cost, but for

---

[1] https://bitbucket.org/apesteam/treelm

the linear system of the acoustic domain, a modal scheme keeps the computational effort per degree of freedom constant over increased spatial orders and solves them efficiently.

The polynomial representation of the DG method also has an advantage in the coupling context. For data exchange at the coupling interface, the polynomial representation can be evaluated at any point on the surface up to the chosen order for the method. In general, the quadrature points of the polynomial on the surface are utilized as exchange points.

3.2 Partitioned Coupling

Partitioned coupling is based on the idea that an entire computational domain can be split into subdomains, where only single physics need to be considered in each subdomain. For the example of fluid-acoustic interaction, this means: we split the whole domain into a subdomain of flow and acoustic generation and a subdomain where only acoustic waves propagate. Small vortices with high energy occur typically around a structure or at high Mach number and generate acoustic waves. In this domain, the small scales of the flow must be resolved. Acoustic waves on the other hand live on larger scales, having less energy, and are transported into the acoustic far field. In this case, the phenomena have to be resolved over long distances. The interactions are realized by a surface coupling between the compressible fluid domain and the acoustic far field. To realize a full coupling which means including information travelling between both subdomains, a bidirectional coupling is deployed, i.e. both domains provide and receive data at the interface.

Figure 1 shows a partitioned coupling example using implicit coupling between structure and fluid and explicit coupling between fluid and acoustic subdomains.

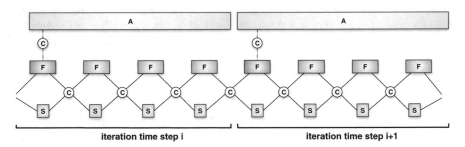

Fig. 1 Overview of parallel execution of the fluid (F) -structure (S) -acoustic (A) simulations. Implicit coupling (C) for fluid-structure interaction presenting an iterative method and explicit coupling (C) at the beginning of the timestep

When coupling several solvers, there are major tasks involved in a coupled setup:

- Steering of the individual single physic solvers
- Data interpolation between non-matching exchange points
- Communication of primitive variables at the interface

These tasks should be handled efficiently in parallel by the coupling tool.

Steering of individual solvers

To control the simulation and the correct update of information at the coupling interface in time, the coupling device should steer the individual solver. The major challenge here is the definition of the synchronization time step.

Data interpolation

For a general setup, allowing individual resolution in each subdomain, the exchange points at the interface do not require to coincide. A non-matching coupling mesh at the interface can occur when e.g. coupling of a higher order Discontinuous Galerkin method which requires information at non-equidistant quadrature points. Figure 2 gives an example of such a non-matching coupling interface, when coupling the same grid resolution but an 8th order Discontinuous Galerkin scheme with a 4th Discontinuous Galerkin scheme where both yield 16 points at a 1d surface. Therefore, an efficient interpolation method is required to transfer the primitive variables of one coupling interface to the other.

Communication

The exchange of data between the solvers is also a task of the coupling device. We aim for large scale simulations on massive parallel systems. Therefore, direct MPI

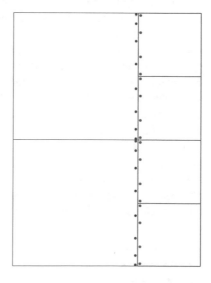

Fig. 2 Example of non-matching exchange points at the coupling interface when coupling the same grid resolution but a 8th order Discontinuous Galerkin scheme (*red*) with a 4th order Discontinuous Galerkin scheme (*blue*) yielding both 16 points at a 1d surface

communication between processes that host coupling elements is essential. This communication takes place at each synchronization time step.

In our approach, explicit coupling is exploited. For the first attempt, we do not allow for adaptive time stepping and sub-cycling of one solver to avoid non-consistent coupling in time. Hence, all subdomains use the same timestep limited by the CFL condition of the explicit timestepping within the solver. This is clearly a lack of ideal performance and it is part of future work. Assuming no adaptive time stepping and a fixed coupling interface, a static load balancing based on heuristics can be achieved by choosing an appropriate number of processes for each subdomain, such that solving each domain takes approximatively the same computational time.

3.2.1 Multi-solver Approach Using Coupling Library *preCICE*

For the multi-solver approach, the focus is on using the solvers as 'black box' which means that the solvers are accessible only via their interfaces for input and output values. Therefore, the aforementioned major tasks of the coupling device are more challenging: Steering between individual solvers, communication of data between executables and accurate interpolation methods between non-matching interfaces. The open-source coupling library *preCICE*[2] offers methods for all these building blocks while allowing for a minimally invasive integration into existing solvers [1]. Additionally, for implicit coupling, which is not part of this paper but a key benefit of *preCICE*, efficient solvers for fixed-point equations derived from coupling conditions are implemented in *preCICE*. Clearly, the major tasks of the coupling device need to work efficiently and should be scalable for distributed data. In [2, 7] development and achievements of *preCICE* working on distributed data are presented.

In *preCICE*, the initialization of the communication is done via exchanging the entire coupling interface via master processes. The communication of coupling participant and coupling library during the time loop is done with point-to-point communication realized via TCP/IP (based on `Boost.Asio`[3]). Coupling different numerical resolution in space requires data at position on the interface which might not be provided by one participant as illustrated in Fig. 2. Therefor interpolation methods between non-matching coupling meshes are required. *preCICE* provides two standard interpolation methods: low order projection-based mapping (nearest neighbor, nearest projection) and second order radial basis function mapping. Both mappings work on pure geometric information assuming 'black box' solvers.

Flexibility is the key benefit of using a coupling tool like *preCICE*. The application programming interface (API) is concise and enables an easy coupling of individual solvers. Additionally, it implements several sophisticated coupling methods, which are required to improve numerical stability at the coupling interface. The advantages are only clouded by the decrease in performance due to generality of a 'black-box' approach.

---

[2] www.precice.org

[3] www.boost.org

Furthermore, the handling of a coupled simulation involves several executables. Porting software, establishing the correct pinning of MPI ranks in this setup, and compiling the job script on a supercomputer is more challenging compared to running a single application.

3.2.2   Integrated Approach Using *APESmate*

The integrated coupling approach *APESmate* is fully implemented in the previously presented framework *APES* [6, 10]. Here, finding a synchronization time step is similar to using the multi-solver approach but the steering of the coupled simulation is direct by accessing the data structure explicitly instead of providing and returning information from a library. Also, communication can be done in a direct way: all components are implemented in a single application which efficiently distributes domains across several processes. Starting with a global communicator, each sub-domain gets its own MPI sub-communicator for domain-internal communications. Therefore, a global communicator is used only for domain-domain communication. During the initialization step, all coupling requests of one subdomain are locally gathered such that only one large communication is necessary instead of multiple small ones. Then this information is exchanged in a round robin fashion. Since every solver in *APES* is based on an Octree data structure and uses a space-filling curve for partitioning, it is easy to get information about the location of the individual exchange points. The identified ranks which accommodate exchange points are provided to the requested domains and these ranks are then used to build communication buffers for data exchange between domains. Point coordinates are only exchanged at the initial step, the point values are evaluated and exchanged via the global communicator once within every time step.

Within the integrated coupling, the application can access solver specific data. Tethering high order DG solver *Ateles*, obtaining data at arbitrary exchange points on the coupling interface can be done via direct evaluation of the polynomial representations. Hence, coupling non-matching grids with different numerical resolution, as shown in Fig. 2, does not involve additional interpolation. This is a key benefit compared to using a multi-solver approach. In the case of coupling other solvers within *APESmate*, e.g. a Lattice-Boltzmann scheme which does not provide a polynomial representation of the solution, the solver is required to provide an interpolation method using its data representation and mathematical formulation. i.e. even if interpolation is necessary, it is done by the data-providing solver, making use of all the knowledge regarding its data and data structure.

In general, *APESmate* is implemented in a way such that surface as well as volume coupling can be realized to increase the range of applications, e.g. the coupling of multi component flow and the electro-dynamic field [5].

Naturally, with this integrated approach, we can only couple solver and methods which are included within *APES* and operating on the underlying data structure of the common mesh library *TreElM*. Up to now, only explicit coupling via data exchange at every time step is available in *APESmate*. However, for fluid-acoustic interaction addressed in this paper, single physics solvers with explicit time step are sufficient.

The performance benefits of *APESmate* as a single application is superior to multi-solver approach due to communications over global MPI communicator and direct control over *Ateles* solver. With respect to load balancing, assuming no adaptive time stepping and not changing the coupling interface, the same static load balancing based on heuristic as presented for the multi-solver approach can be applied. Also, a dynamic load balancing can be deployed easier. From the user perspective, the handling of an integrated approach with one executable is facilitated.

4 Results

In this section, we show the comparison of the two presented coupling approaches, using the external library *preCICE* as well as the integrated approach *APESmate*. We setup two different scenarios, one coupling the same equations system on both sides, but using different mesh sizes and approximation orders, and the other coupling different equation systems, on the same and on different meshes and orders. When using *preCICE*, we also vary the interpolation method between first order nearest neighbor interpolation and second order radial basis functions.

The second part of this section describes the performance scalability of both established coupling strategies on modern supercomputer.

4.1   Simulation Setup

We show two dedicated test cases:

(a)   Gaussian distribution in density on a 2-dimensional domain (Fig. 3a)
(b)   Gaussian distribution pressure on a 3-dimensional domain (Fig. 3b)

Testcase (a) is used for coupling the same equations systems and (b) to couple two different equations e.g. a non-linear flow subdomain with a linearized Euler domain.

For test case (a) we will refer in the following to as left and right subdomain as illustrated, and for test case (b) as flow domain and acoustic domain.

4.2   Numerical Results

4.2.1   Bidirectional Coupling of the Same Equations Systems: Flow with Flow

To test the coupling of twice the same equation systems, we deploy a 2-dimensional Gaussian density distribution which travels from left to right due to advection of the flow in positive x-direction, see Fig. 3a. The whole domain is a two dimensional

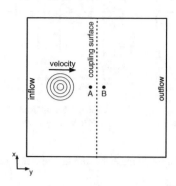

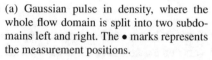

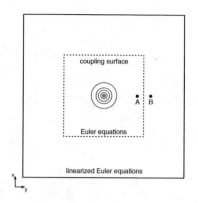

(a) Gaussian pulse in density, where the whole flow domain is split into two subdomains left and right. The • marks represents the measurement positions.

(b) Gaussian pulse in pressure, where the inner box is the flow domain (Euler equations) and the surrounding is the acoustic domain (linearized Euler equations). The • marks represents the measurement positions.

Fig. 3 Sketch of the two dedicated simulation setups

4×4 xy-plane, which is split into a left and a right subdomain. As described in Sect. 2, for the Euler equations (1)–(3) the ideal gas is considered. Here, the isentropic coefficient is chosen to be $\gamma = 1.4$ and the ideal gas constant is $R = 296.0$. The density is initially given as a Gaussian pulse shifted by $(x_0 = -1.0, y_0 = 0.0)$ to be fully located in the left subdomain:

$$\rho = \rho_0 + \rho_{pulse} \cdot \exp\left(-[(x + x_0)^2 + (y + y_0)^2]/d \cdot \log(2)\right)$$

with background density $\rho_0 = 1.0$, amplitude of the pulse $\rho_{pulse} = 1.0$ and half width of the pulse $d = 0.02$. The flow is initialized with a constant velocity field, $\mathbf{v}_{t=0} = [2.0\ 0.0]^T$ and pressure, $p_{t=0} = 8.0$. As shown in Fig. 3a, the left and right boundary conditions are inflow and outflow respectively, whereas the upper and lower boundaries are set to the full state $[\rho_0, \mathbf{v}_0, p_0]$. The analytical solution of the density pulse traveling through the flow at time t is

$$\rho_{ref} = \rho_0 + \rho_{pulse} \cdot \exp\left(-[(x + x_t)^2 + (y + y_t)^2]/d \cdot \log(2)\right) \tag{7}$$

with the location of the pulse $x_t = v_{0x}t - x_0$, $y_t = v_{0y}t - y_0$.

4.2.2 Results and Comparison

The investigation is done for both established methods, the integrated approach *APESmate* and the multi-solver approaches using *preCICE*. The two approaches differ mainly in the way they obtain the data required for the one side from the data

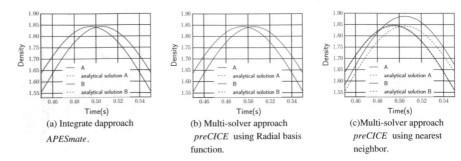

Fig. 4 Comparison of numerical and analytical result in both subdomains for coupling of different resolutions: *Left* $\mathscr{L} = 4$, $\mathscr{O} = 16$; *Right* $\mathscr{L} = 4$, $\mathscr{O} = 22$

provided by the other. When looking at Fig. 2, the left domain delivers and expects data on the red points, while the right side delivers and expects data at the blue points. When using the multi-solver approach, the coupling tool *preCICE* interpolates from red to blue and from blue to red points. Two different interpolation methods, nearest neighbor and radial basis function, are available in *preCICE*. The integrated approach in *APESmate*—as it is able to access directly the high-order polynomials within the left as well as within the right part of the domain—directly evaluates the polynomials at the points requested and thus does not interpolate at all.

For validation purposes, a first run of the simulation is performed using the same mesh and the same order of the DG scheme on both sides. Thus, the data exchange points match on both sides, and interpolation reduces to pure injection. The two different interpolation schemes which are tested against each other within the multi-solver approach with *preCICE* should not show any difference, neither compared to each other nor compared to the integrated approach with *APESmate*. For this pre-testcase, all results coincide as expected.

The next variation now checks the influence of the interpolation in the case of non-matching grids as in Fig. 2. Non-matching grids are obtained when using different mesh sizes or different approximation orders in the DG scheme. We refer to the grid resolution as refinement Level $\mathscr{L}$ of the Octree mesh and $\mathscr{O}$ for the numerical order of the DG scheme in space.

Figure 4 shows the comparison of the different coupling strategies when coupling two different discretizations, i.e. left: $\mathscr{L} = 4$, $\mathscr{O} = 16$; right: $\mathscr{L} = 4$, $\mathscr{O} = 22$. It is measured at positions A and B (Fig. 3a) at the point in time when the maximum amplitude of density pulse is reached. The integrated approach (Fig. 4a) as well as the multi-solver approach using second order radial basis functions for data interpolation at the exchange points (Fig. 4b) give good results and are identical with the analytical solution. The first order nearest neighbor interpolation in the multi-solver approach (Fig. 4c) produces an overshooting in point B (to the right of the coupling interface), compared to the solution in point A (left to the coupling interface), and the analytical solution.

4.2.3 Error Analysis

For the error analysis, we compare the simulation result to the analytical solution and present the error, i.e. the difference between analytical solution and simulation at the maximum density of the Gaussian pulse in both subdomains. Table 1 gives an overview of the simulation error at points A (left to the coupling interface) and B (right to the coupling interface).

As mentioned before, testing the same numerical resolution in both subdomains ensures matching exchange points at the coupling interface and avoids influences

Table 1 Comparison of simulation error at maximum of the Gaussian distribution at points ± 0.01 distance from the coupling surface

Left domain			Right domain		
$\mathscr{L}$	$\mathcal{O}$	Error	$\mathscr{L}$	$\mathcal{O}$	Error
(a) Integrated approach APESmate					
4	16	−2.3833e-4	4	8	1.8816e-4
4	16	2.7661e-4	4	12	−7.9379e-4
4	16	5.7715e-5	4	16	4.1064e-5
4	16	1.1531e-5	4	22	4.5484e-6
4	16	−1.7762e-5	4	32	−2.0103e-5
4	16	−2.1650e-5	3	32	−1.8538e-5
4	16	1.1151e-4	5	8	−9.2129e-4
3	64	−2.9949e-6	4	32	−3.5793e-7
(b) Multi-solver approach with preCICEusing 2nd order radial Basis function interpolation					
4	16	−2.7129e-3	4	8	1.9446e-3
4	16	2.820e-4	4	12	1.0909e-3
4	16	5.7715e-5	4	16	4.1064e-5
4	16	4.0299e-5	4	22	−3.1967e-4
4	16	−1.4563e-5	4	32	−3.0865e-4
4	16	3.4437e-5	3	32	3.3906e-4
4	16	1.4550e-4	5	8	1.0778e-3
3	64	−7.6588e-6	4	32	−2.6305e-5
(c) Multi-solver approach with preCICEusing Nearest Neighbor interpolation					
4	16	1.5106e-2	4	8	−1.3613e-1
4	16	1.3116e-2	4	12	2.7013e-3
4	16	5.7715e-5	4	16	4.1064e-5
4	16	3.6480e-3	4	22	4.3454e-2
4	16	5.6850e-3	4	32	−3.7787e-2
4	16	5.0766e-4	3	32	3.6137e-2
4	16	7.2353e-4	5	8	8.0711e-4
3	64	−8.3883e-4	4	32	2.3469e-3

of the data interpolation required for the multi-solver approach. Comparing the simulation error for the same numerical resolution, i.e Left: $\mathscr{L} = 4$, $\mathscr{O}(16)$; Right: $\mathscr{L} = 4$, $\mathscr{O}(16)$, 3 line in Table 1, demonstrates a good agreement of all strategies as expected.

Comparing different numerical resolutions demonstrates the important influence of the data mapping strategy on the coupling interface. Table 1 indicates that a first order nearest neighbor interpolation is no option for coupling non-matching grids since disproportionally large numerical errors arise. Only one out of the 8 combinations results in errors comparable to the other approaches, which is line 7, Left: $\mathscr{L} = 4$, $\mathscr{O}(16)$; Right: $\mathscr{L} = 5$, $\mathscr{O}(8)$. This might be due to the exact position of the exchange points on the surface. In this setting, the distance between the coupling points on both sides of the coupling interface is nearly minimal.

In general, it can be stated that the integrated *APESmate* as well as the multi-solver approach *preCICE* with a second order radial basis function (RBF) interpolation give good results, whereby the simulation error for *APESmate* yields a simulation error which is even one or two orders of magnitude lower than for the *preCICE* and, order RBF approach.

4.2.4 Performance Results

In this section, we present the performance of integrated coupling *APESmate* and multi-solver coupling *preCICE* using nearest neighbor interpolation only on the SuperMuc Phase 1 IBM system at LRZ, Munich. This system comprises a total of 9216 nodes on 18 islands with 2 Sandy Bridge-EP Xeon E5-2680 processor with 8 cores per node resulting in 147,456 cores. The nodes are connected with Infiniband FDR10. For performance measurements, both coupling approaches are scaled up to a single island, i.e 512 compute nodes or 8192 cores. Using more than one island is not possible at the moment due to limitations of MPI-IO on SuperMuc. Only MPI parallelism is considered here.

A 3D version of the test case presented in the previous section (Gaussian pulse in density traveling from left to right domain) is used with a total problem size of 8192 elements, i.e 4096 elements per coupling domain. The total problem size of 8192 elements is chosen such that there is at least one element per core and the polynomial order, $\mathscr{O}$ is chosen to fit maximally to the memory per node which is found to be $\mathscr{O}(20)$. The simulations are run for 100 iterations. The number of degrees of freedom per element is 109760, resulting in 163840000 *DoF* for problem size of 4096 elements per domain.

In Fig. 5, the strong scaling of both strategies *APESmate* and *preCICE* coupling on left and right domain are shown together with the number of processes on the X-axis and the total run time in seconds on Y-axis. Both approaches have good scalability up to 1024 processes per domain. Beyond that, the integrated approach *APESmate* does not scale anymore and gets flat where as the scalability of the multi-solver approach *preCICE* gets worse, i.e run time increases with number of processes. The increase in run time might be due to load imbalances stemming from only few processes

Fig. 5 Strong scaling of integrated coupling *APESmate* and multi-solver coupling *preCICE*

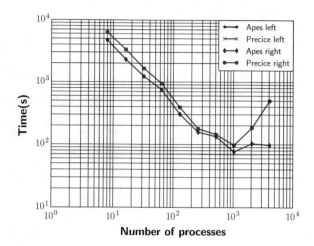

participating in the coupling. Nevertheless, at all points, the integrated approach *APESmate* is roughly 20 % faster then the multi-solver approach using the external coupling library *preCICE*. To determine the performance critical step, the overall run time is split into initialization, computation and coupling. Figure 6 shows this breakdown of overall run time for *APESmate* in Fig. 6a and *preCICE* in Fig. 6b. For both approaches, the initialization step is further split into initialization of solver and coupling. In Fig. 6, the total initialization time increases with the number of processes for both approaches but *preCICE* initialization time is much higher than *APESmate*. As stated in [2], the initialization step is not fully parallelized and work in progress. In *APESmate*, we can measure the time spent on computation and initialization separately where as in *preCICE* coupling initialization time is part of computation time and difficult to calculate explicitly. This can be seen from the computation time in Fig. 6a since both approaches uses the *Ateles* solver which is scalable on its own. The coupling step involving the evaluation of point values and data exchange is faster with *preCICE* than *APESmate*. The coupling in *APESmate* involves the evaluation point values using polynomial which is expensive but more accurate than the fast, but inaccurate nearest neighbor approach used in *preCICE*. Also, the *APESmate* coupling approach shows better scalability than *preCICE*. From Fig. 6b, we can conclude that for *preCICE*, the increase in run time beyond 1024 processes per domain is mainly due to the initialization of coupling.

4.2.5   Bidirectional Coupling of Differing Equation Systems: Euler with Linearized Euler

To test the coupling of differing equation systems e.g. Euler equations with linearized Euler equations, we use an acoustic pulse initialized at time t = 0 with a Gaussian pressure distribution which is spreading spherically symmetric with respect to the

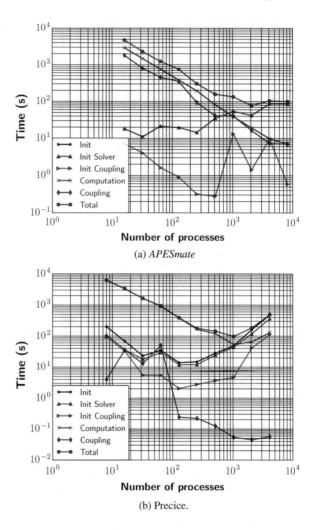

Fig. 6 Strong scaling breakdown of overall time into initialization, computation and coupling of integrated approach *APESmate* and multisolver approach *preCICE*

(a) *APESmate*

(b) Precice.

origin of the pulse as described in [8], sketched in Fig. 3. The 3-dimensional flow domain in which the pulse is located is a $20 \times 20 \times 20$ box with a surrounding acoustic domain of size $60 \times 60 \times 60$. For both domains, the isentropic coefficient is set to $\gamma = 1.4$ and the ideal gas constant is $R = 296.0$. Additionally, for the acoustic domain, treating the linearized Euler equations (4)–(6), the background flow is set to $\rho_0 = 1.0$, $\mathbf{v_0} = \begin{bmatrix} 0.0, 0.0, 0.0 \end{bmatrix}^T$, $p_0 = \frac{1}{\gamma}$ yielding a speed of sound $c = 1.0$. For the inner flow domain, the initial condition for the Euler domain is a Gaussian pressure distribution:

$$p = p_0 + p_{pulse} \cdot \exp\left(-[(x + x_0)^2 + (y + y_0)^2 + (z + z_0)^2]/d \cdot \log(2)\right)$$

with amplitude of the pulse $p_{pulse} = 0.001$ and half width set to $d = 3$. The background for the flow is set to background of the acoustic domain. The Euler domain is initialized with density $\rho_{t=0} = 1.0$ and velocity $\mathbf{v}_{t=0} = [0.0, 0.0, 0.0]^T$. For the surrounding acoustic domain, the initial condition $[\rho^a, \mathbf{v}^a, p^a]^T$ is specified to 0, since at the start of the simulation, no acoustic perturbation should occur. The outer boundaries for the acoustic domain are set to a Dirichlet boundary condition for all state variables i.e. $\rho^a = 0.0$, $\mathbf{v}^a = [0.0, 0.0, 0.0]^T$, $p^a = 0.0$. The analytical solution for a Gaussian pressure distribution spreading spherically symmetric with respect to the origin $(0.0, 0.0, 0.0)$ and the radial distance $r = \sqrt{(x - x_0)^2 + (y - y_0)^2 + (z - z_0)^2}$ is :

$$p = p_0 + p_{pulse} \cdot \left[\frac{r-c \cdot t}{2 \cdot r} \cdot \exp\left(-\log(2) \cdot \left(\frac{r-c \cdot t}{b}\right)^2\right) \right.$$
$$\left. + \frac{r+c \cdot t}{2 \cdot r} \cdot \exp\left(-\log(2) \cdot \left(\frac{r+c \cdot t}{b}\right)^2\right) \right] \qquad (8)$$

with speed of sound defined by the material $c = \sqrt{\frac{\gamma \cdot p_0}{\rho_0}}$.

4.2.6 Results and Comparison

We look at the temporal evolution at the specific points A and B close (± 0.01) to the coupling interface as sketched in Fig. 3a. Figure 7 shows a coupled simulation using the non-linear Euler equations in the inner domain, and the linearized Euler

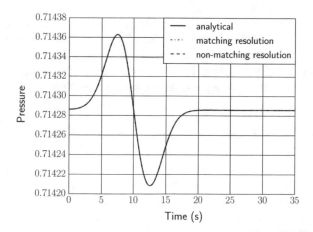

Fig. 7 Time evolution of pressure at measure positions, see Fig. 3b. Comparing a coupled simulation with same numerical resolution—matching [flow domain: 8,000 elements, element size = 1, $\mathcal{O}(6)$, acoustic domain: 208,000 elements, element size = 1, $\mathcal{O}(6)$] with different numerical resolution—non-matching [flow domain: 8,000 elements, element size = 1, $\mathcal{O}(6)$, acoustic domain: 3,250 elements, element size = 5, $\mathcal{O}(12)$]

Table 2 Absolute and relative error for the different simulations of the Gaussian pulse distribution measured at positions ±0.1 off from the coupling surface in the flow respectively in the acoustic domain. Relative error is normalized to the acoustic perturbation in the pressure since this is the travelling information

	Flow domain		Acoustic domain	
	Absolute error	Relative error	Absolute error	Relative error
Maximum of pressure distribution				
Matching coupling	1.5e-07	1.9e-03	7.25e-08	9.37e-04
Non-matching coupling	1.53e-07	1.98e-03	8.07e-08	1.04e-03
Minimum of pressure distribution				
Matching coupling	1.14e-07	1.471e-03	3.73e-08	4.83e-04
Non-matching coupling	1.16e-07	1.497e-03	4.27e-08	5.53e-04

equations in the outer domain. In the first setup, only the equations are switched, but mesh level and order of the scheme are kept the same in both parts of the domain (matching resolution). In a second setup, the effort for the outer domain is decreased by using a coarser mesh, but a higher order in the DG scheme compared to the inner domain (non-matching). The numerical configuration for the matching simulation is flow domain: 8,000 elements, element size = 1, $\mathcal{O}(6)$, acoustic domain: 208,000 elements, element size = 1, $\mathcal{O}(6)$. The configuration for the non-matching setup is flow domain: 8,000 elements, element size = 1, $\mathcal{O}(6)$, acoustic domain:3,250 elements, element size = 5, $\mathcal{O}(12)$.

Table 2 shows the comparison of the results for matching and non-matching grid in terms of absolute and relative error in pressure. The table shows the good accordance of the results, i.e. the quality of the solution is the same for the finer mesh with lower order (matching configuration) as for the coarser mesh with higher resolution. We will now investigate the gain in performance by this variation.

4.2.7 Performance Benefits for Coupling Differing Equation Systems

Coupling non-linear Euler equations with linearized Euler equations yield different computational load on the corresponding subdomains. Therefore, a coupled simulation with properly chosen computational resources can reduce the computational cost. Besides the variation in the numerical parameters (matching/non-matching), also the distribution of the two subdomains to available compute resources can be optimized. Table 3 illustrates the overall runtime of the Gaussian pressure pulse simulation for different settings of the parallel distribution where the overall number of 512 MPI-ranks is used on the SuperMuc Phase 1 IBM system at LRZ, Munich. The particular number of MPI ranks is chosen to fill a full node with 16 processes. In the

Table 3 Load balancing for different distributions of total 512 MPI-ranks on SuperMUC of a coupled simulation of Gaussian pressure pulse against monolithic simulation. Using 64 ranks for the flow domain and 448 for the acoustic domain yield the faster computation for same resolution on both subdomains

Type	Number of MPI-ranks		Total computation time (s)
	Flow domain	Acoustic domain	
Matching coupling	256	256	1336
	128	384	886.65
	112	400	845.573
	96	416	824.695
	80	432	847.923
	64	448	832.06
	48	464	838.607
	32	480	902.342
	16	496	1535
Non-matching coupling	96	416	476.886

case of matching grids, we found that utilizing 96 ranks for the flow domain and 416 for the acoustic domain yields the fastest computation with 824s. Table 3 illustrates how the imbalance moves from the acoustic domain, when using a low number of MPI-ranks here, to the flow domain since when using more MPI-ranks in acoustics, less in the flow domain. Using a too small number of MPI-ranks for this domain, i.e. 16 MPI ranks for the flow domain as well using a too small number of MPI-ranks for the acoustics, leads to the longest runtimes of 1336 and 1535 s respectively. The optimal setting is neither to the one nor to the other extreme, but in a medium, best suited distribution which reduces the runtime by roughly a factor of 2.

The best setup nevertheless is the adapted configuration which uses a much coarser mesh for the acoustic domain, yet settled by an increased order for the DG scheme. By this change in the numerical parameters, the runtime can be decreased by another factor of 2. These performance benefits become even more crucial when enlarging the acoustic far field even more or when solving different numerical problem where the length scale in the flow domain are magnitudes smaller than in the presented testcase.

5 Conclusion

Partitioned coupling is a promising strategy to solve multi-scale as well as multi-physic problems on todays supercomputer. By splitting the whole domain into single physics subdomains and enabling interaction via surface coupling, each single physics domain can be solved by individual solvers using numerical methods which are perfectly tailored to the underlying physics. Hence, problems which might not be not feasible in a monolithic approach, due to e.g. too different length scales ending in large computational costs, can be accomplished.

We presented two different coupling approaches namely a multi-solver approach utilizing an external coupling library which takes care about steering, data mapping as well as data communication, but uses the individual solvers as black boxes, and an integrated approach, making use of all knowledge available on the solver, implemented within one numerical framework. This approach suffers from a loss of generality but gains performance.

Exploiting a higher order method in the solver has the advantage that polynomial approximations on the coupling surface are available and therefore, can be used within the integrated coupling approach *APESmate*. In contrast, the multi-solver approach with the coupling library *preCICE* requires an additional interpolation method for the data mapping. For non-matching grids, which typically occur when coupling different numerical resolution, using first order interpolation show unsatisfactory simulation error. Using direct polynomial evaluation for the data mapping, which is one key benefit of *APESmate* exhibits very good results when coupling high order which was shown with the example of coupling 64th order in space with and 32th order in space. For medium order, using *preCICE* with 2nd order radial basis functions and *APESmate* with direct evaluation, both yield satisfying numerical results.

Comparing the performance of the integrated approach *APESmate* with the multi-solver approach using *preCICE* on a modern supercomputer SuperMuc at LRZ, Munich, *APESmate* shows an advantage of 20 % lower overall computation time. This confirms the expected performance benefits gained by the tight integration of the coupling with the solver, which allows for exploitation of knowledge about internal data structures. But even that the multi-solver approach can not compete with a fully integrated approach in terms of overall runtime, the scalability is nevertheless satisfying.

Partitioned coupling leads to different work load in the single physics domains and hence, properly chosen number of compute resources can reduce the overall computational costs. This is shown on the example of a Gaussian pressure distribution, where a non-linear flow domain (Euler equations) is coupled to a surrounding acoustic domain (linearized Euler equations). Chosing the right distribution of MPI-ranks per subdomain, the computational cost is reduced by a factor of 2. Adaption of the numerical resolution in the individual domains, e.g. by coarsening the grid resolution and increasing the order in the acoustic domain can reduce the computational cost even more, in our example by a factor of 2 compared to the matching resolution coupling.

The focus of future work is on numerical challenges, in particular the coupling of different timesteps. Enabling subcycling of one solver by assuring a consistent timestep even for large differences in individual timesteps will give further performance benefits.

Acknowledgements The financial support of the priority program 1648 - Software for Exascale Computing 214 (www.sppexa.de) of the German Research Foundation. The performance measurements were performed on the Supermuc supercomputer at Leibniz Rechenzentrum (LRZ) der Bayerischen Akademie der Wissenschaften. The authors wish to thank for the computing time and the technical support.

References

1. Bungartz, H.J., Lindner, F., Gatzhammer, B., Mehl, M., Scheufele, K., Shukaev, A., Uekermann, B.: Precice – a fully parallel library for multi-physics surface coupling. Comput. Fluids (2015). Accepted
2. Bungartz, H.J., Lindner, F., Mehl, M., Scheufele, K., Shukaev, A., Uekermann, B.: Partitioned Fluid-structure-acoustics interaction on distributed data - coupling via preCICE. In: Bungartz, H.J., Neumann, P., Nagel, E.W. (eds.) Software for Exa-scale Computing - SPPEXA 2013–2015. Springer, Berlin (2016)
3. Hesthaven, J.S., Warburton, T.: Nodal Discontinuous Galerkin Methods: Algorithms, Analysis, and Applications, 1st edn. Springer Publishing Company, Incorporated (2007)
4. Klimach, H.G., Hasert, M., Zudrop, J., Roller, S.P.: Distributed octree mesh infrastructure for flow simulations. In: J. Eberhardsteiner (ed.) ECCOMAS 2012 - European Congress on Computational Methods in Applied Sciences and Engineering, e-Book Full Papers (2012)
5. Masilamani, K., Klimach, H., Roller, S.: Highly efficient integrated simulation of electro-membrane processes for desalination of sea water. In: W.E. Nagel, D.B. Kröner, M.M. Resch (eds.) High Performance Computing in Science and Engineering '13, pp. 493–508. Springer, New York (2013). doi:10.1007/978-3-319-02165-2
6. Roller, S., Bernsdorf, J., Klimach, H., Hasert, M., Harlacher, D., Cakircali, M., Zimny, S., Masilamani, K., Didinger, L., Zudrop, J.: An adaptable simulation framework based on a linearized octree. In: Resch, M., Wang, X., Bez, W., Focht, E., Kobayashi, H., Roller, S. (eds.) High Performance Computing on Vector Systems 2011, pp. 93–105. Springer, Berlin (2012)
7. Shukaev, A.K.: A fully parallel process-to-process intercommunication technique for precice. Master's thesis, Institut für Informatik, Technische Universität München (2015)
8. Tam, C.K.W.: Computational Aeroacoustics. Cambridge University Press (2012). http://dx.doi.org/10.1017/CBO9780511802065. Cambridge Books Online
9. Zudrop, J.: Efficient numerical methods for fluid- and electro-dynamics on massively parallel systems. Ph.D. thesis, RWTH Aachen (2015)
10. Zudrop, J., Klimach, H., Hasert, M., Masilamani, K., Roller, S.: A fully distributed CFD framework for massively parallel systems. In: Cray User Group 2012. Stuttgart, Germany (2012)

The Spectral Structure of a Nonlinear Operator and Its Approximation II

Uwe Küster

Abstract Linear operators allow for a structural analysis by their spectra and the related decomposition in stable linear subspaces. This does not apply to nonlinear operators, which are relevant for most natural phenomena. But this problem could be overcome. The nonlinear operator can induce a special linear operator in a large linear space. The induced linear operator is named Koopman operator offering structures to be mapped to the original settings. We try to give approaches for a numerical handling of some properties, generalizing the approach of the Dynamic Mode Decomposition of Peter Schmid.

1 Introduction

This paper is directly related to a first part [9] from the author and is to be considered as an extension of the numerical approaches. The analysis of linear operators as used in simulation problems is attractive because of the inherent spectral structures, which might simplify the theoretical and numerical analysis or may be substantial part of the observable natural properties. Surely more important are the nonlinear operators governing nearly all natural processes. They are typically very complex and seem to lack the obvious structures induced by linearity. Sometimes people try to linearize the operators in a small region of interest to get linear structures. But this approach does not show global implications of local solutions, might show critical features, but not there global effects. But there is a functional analytic mechanism, by which an nonlinear operator with some weak assumptions can be extended to a linear operator on another larger space. This so-called Koopman operator has structures which might be helpful for analysis of the natural problem. The Koopman operator appears in Ergodic Theory founded by Ludwig Boltzmann and developed later by John von Neumann, John von Neumann, George David Birkhoff, Bernard Osgood Koopman, Norbert Wiener, Aurel Friedrich Wintner. See the monograph [1] also

U. Küster (✉)
High Performance Computing Center Stuttgart (HLRS), Nobelstraße,
19 70569 Stuttgart, Germany
e-mail: kuester@hlrs.de

© Springer International Publishing AG 2016   83
M.M. Resch et al. (eds.), *Sustained Simulation Performance 2016*,
DOI 10.1007/978-3-319-46735-1_7

for modern developments. On the other hand their are more empirically motivated developments as the Dynamic Mode Decomposition of Peter Schmid [7]. These showed to be related to the Koopman operator theory, as pointed out by Igor Mesić and coworkers in [4] and also Clarence Rowley and his coworkers in [5].

Our interest here is to give some remarks on the relation of the spectral structure of the Koopman operator to the possibility to approximate a part of the spectrum numerically. The approach is motivated by the Dynamic Mode Decomposition and generalizing it. No examples are given.

2   The Koopman Operator

Let

$$\varphi : K \longrightarrow K \tag{1}$$

be a continuous nonlinear operator on the compact space K and assume $\mathscr{F}$ being a linear subspace of "observables", of $C(K)$ the continuous functions on K. $\mathscr{F}$ shall have the stability property

$$f \in \mathscr{F} \Rightarrow f \circ \varphi \in \mathscr{F} \tag{2}$$

that means, that an observable coupled with the operator is again in the observable space. Observables might be any useful functional on the space of interest as the mean pressure of a (restricted) fluid domain Ω or the evaluation operators δ_x at all points $x \in \Omega$. The nonlinear operator φ has no further restrictions. It might describe non wellposed unsteady problems, the case where trajectories are not convergent (also strange attractors), chaotic or turbulent behaviour, mixing fluids, particle systems or ensembles of trajectories for weather forecast. The operator could also be defined by an agent based system for the simulation of traffic, epidemics, social dependencies, where the agents determine their next status by the current status of some other neighbouring agents. In this case K is the set product of the status of all agents with some topology and surely not a subset of a vector space in contrast to $\mathscr{F}$. Even the pictures of a movie could be understood as elements of a space K ordered in a trajectory which is produced by an unknown operator φ (Kutz, SIAM Conference on Applications of Dynamical Systems 2015, Snowbird). As an important numerical example φ could be taken as a time discretization of the Navier-Stokes equations on a finite set of grid points in a domain and time steps. It is even possible to understand K here as the product of the status of all variables on the discretization grid together with varying boundary conditions and geometrical parameters.

By a simple mechanism the nonlinear operator φ acting on a set without linear structure induces a linear operator on the space of observables $\mathscr{F}$. The operator T_φ on the observables defined by

$$T_\varphi : \mathscr{F} \longrightarrow \mathscr{F} \tag{3}$$

$$f \mapsto T_\varphi f = f \circ \varphi \tag{4}$$

is named the Koopman operator of φ on $\mathscr{F}$ [2]. It is immediately clear that T_φ is linear and continuous.

As an infinite dimensional operator T_φ may have a (complicated) spectrum with discrete and continuous parts. We are mainly interested in the point spectrum with eigenvalues providing eigenfunctions which are elements of $\mathscr{F}$.

The drawback is that T_φ acts on an infinite dimensional space even in very simple cases, which is a pain for numerical analysis.

2.1 Unusual properties of the Koopman-Operator

Even for simple cases the space of observables $\mathscr{F}$ is large because of the stability property. For numerical applications it can be restricted in a still meaningful way. Only a small finite part of the point spectrum can be approximated in a numerical way. The eigenvectors or eigenfunctions f are elements of the space of observables $\mathscr{F}$, not of the state space K as it would be in the linear case. They fulfill **Schröders functional equation**

$$f(\varphi q) = \lambda f(q) \quad \forall q \in K$$

It might be difficult to be interpret this equation in terms of specific phenomena in application cases. The eigenfunctions f are typically nonlinear as K has no linear structure. The point spectrum $P\sigma(T_\varphi)$ of the Koopman operator has unusual properties. Dependent on the extent of the space of observables $\mathscr{F}$ and eigenpairs (λ, f), (λ_1, f_1), (λ_2, f_2) with $f_1 \cdot f_2 \neq 0$ we have the following. $\lambda_1 \cdot \lambda_2 \in P\sigma(T_\varphi)$ with eigenfunction $f_1 \cdot f_2$ and $|\lambda| \in P\sigma(T_\varphi)$ with eigenfunction $|f|$ assuming, that $f_1 \cdot f_2, |f| \in \mathscr{F}$.

3 The Numerical Approach: The Space of Observables

For numerical calculations it is important to find a reasonable linear space of observables $\mathscr{F}$ which should be **as small(!) as possible**, whereas functional analysis takes the continuous functions on K ($\mathscr{F} = C(K)$) or even larger spaces.

Necessary are the stability conditions (2). In a numerical context this can be reached by simply iterating the observable. The smallest reasonable numerical setting is to investigate the finite sequence

$$G^f(q) = \left[g_k^f(q)\right]_{k=0,\cdots,n} = \left[f\left(\varphi^k q\right)\right]_{k=0,\cdots,n} \tag{5}$$

for a single observable f starting with an arbitrary single state $q = q_0 \in K$. Starting with $q' = \varphi^j q$ is also a reasonable option enforcing the significance of a shifted sequence on the same trajectory. Nevertheless the trajectory could be large, even dense in K. A finite number of linear independent observables S can be combined in this way in a vector h. Explicit knowledge of the operator φ is not needed for numerical purposes; the effect of the operator on the state space as measured by the observables is sufficient. This is very helpful for numerical purposes. Essential for the following to find a (normalized) vector c of degree $p \le n$ with the property $G^f(q)\ c \approx 0$ in some way to be defined. This vector c is understood as a polynom coefficient vector with polynomial roots λ_l.

3.1   Relation to the Koopman Operator

The values for the k-times iterated Koopman operator T_φ on the trajectory are

$$\left(T_\varphi^k h\right)(q_l) = h\left(\varphi^k q_l\right) = h\left(\varphi^{k+l} q_0\right) = g_{k+l} \quad \forall\, k,\, l \in \mathbb{N}_0 \tag{6}$$

The space of observables $\mathscr{F}$ and the Koopman operator for the restriction of φ to the trajectory are completely described in this way. The vectors g_{k+l} are given by measurements or numerical calculations. The Krylov space matrix of the first p iterations of the infinite-dimensional vector consisting on finite dimensional subvectors $(h(q_l))_{l \in \mathbb{N}_0}$ is

$$
\begin{bmatrix}
h(q_0) & \left(T_\varphi h\right)(q_0) & \dots & \left(T_\varphi^p h\right)(q_0) \\
h(q_1) & \left(T_\varphi h\right)(q_1) & \dots & \left(T_\varphi^p h\right)(q_1) \\
h(q_2) & \left(T_\varphi h\right)(q_2) & \dots & \left(T_\varphi^p h\right)(q_2) \\
h(q_3) & \left(T_\varphi h\right)(q_3) & \dots & \left(T_\varphi^p h\right)(q_3) \\
& \vdots &
\end{bmatrix}
=
\begin{bmatrix}
h(q_0) & h\left(\varphi^1 q_0\right) & \dots & h\left(\varphi^p q_0\right) \\
h(q_1) & h\left(\varphi^1 q_1\right) & \dots & h\left(\varphi^p q_1\right) \\
h(q_2) & h\left(\varphi^1 q_2\right) & \dots & h\left(\varphi^p q_2\right) \\
h(q_3) & h\left(\varphi^1 q_3\right) & \dots & h\left(\varphi^p q_3\right) \\
& \vdots &
\end{bmatrix}
$$

$$
=
\begin{bmatrix}
g_0 & g_1 & \cdots & g_{0+p} \\
g_1 & g_2 & \cdots & g_{1+p} \\
g_2 & g_3 & \cdots & g_{2+p} \\
g_3 & g_4 & \cdots & g_{3+p} \\
& \vdots &
\end{bmatrix}
\tag{7}
$$

This is a Hankel type matrix. Any row in this infinite matrix is the left shifted row above. The entries are constant along antidiagonals. Only a finite number of rows can be handled in a numerical procedure.

3.2 Hankel Matrices

We therefore consider the Hankel matrix of measurements

$$
G_{0:n-p,\ 0:p} \ = \
\begin{bmatrix}
g_{0+0} & g_{1+0} & \cdots & g_{p+0} \\
g_{0+1} & g_{1+1} & \cdots & g_{p+1} \\
\vdots & \vdots & & \vdots \\
g_{0+n-p} & g_{1+n-p} & \cdots & g_n
\end{bmatrix}
\tag{8}
$$

where the j-th line is given by the measurements $0 + j : p + j$ for $j = 0, \cdots, n - p$. The elements may be whole vectors consisting on observables. The whole matrix may be understood as a set of combined measurements. We try to minimize the l_2-norm of the error r for a coefficient-vector c with deg $c = p$

$$
G\,c = r \tag{9}
$$

with

$$
r =
\begin{bmatrix}
r_0 \\
r_1 \\
\vdots \\
r_{n-p}
\end{bmatrix}
\tag{10}
$$

in some appropriate way. r consists on multiple vectors of element type. The vector c is a polynom coefficient vector of degree deg p and is normalized in the l_2-Norm.

3.3 Convolution of Polynom Coefficient Vectors

Definition 3.1 The **convolution** of two coefficient vectors c with deg $c = p$ and b with deg $b = q$ is the given by the coefficient vector of the product polynom

$$
(c * b)(\lambda) = c(\lambda)\,b(\lambda) \quad \forall \lambda \in \mathbb{C} \tag{11}
$$

The **convolution-matrix** $\mathfrak{A}_n(c)$ for a coefficient vector c of deg $c = p$ is the matrix of the progressively shifted vector c

$$
\mathfrak{A}(c) = \mathfrak{A}_n(c) = \quad
\begin{array}{cc}
 & \begin{array}{cc} 0 & \quad n-p \end{array} \\
\begin{array}{c} 0 \\ 1 \\ \cdot \\ \cdot \\ \cdot \\ p \\ \cdot \\ n \end{array} &
\left(
\begin{array}{cc}
c_0 & \\
c_1 & \ddots \\
 & \quad c_0 \\
 & \quad c_1 \\
c_p & \quad \vdots \\
 & \ddots \\
 & \quad c_p
\end{array}
\right)
\end{array}
\tag{12}
$$

The convolution and the convolution matrix have the following properties.

Theorem 3.2 *Let c be the coefficient-vector of a polynom with* $\deg c = p \ (c_p \neq 0)$.

1. *The convolution matrix acts as the convolution product* $(\deg b = n - p)$

$$
\mathfrak{A}(c)\, b = c * b \tag{13}
$$

2. *The convolution matrix has full rank*

$$
\dim \operatorname{Im} \mathfrak{A}_n(c) = n - p + 1 \tag{14}
$$

3. *If* $\deg b = q \leq n - p$ *for a second polynom* b, *then for the product of the two polynoms we have*

$$
\mathfrak{A}_n(c * b) = \mathfrak{A}_n(c)\,\mathfrak{A}_{n-p}(b) = \mathfrak{A}_n(b)\,\mathfrak{A}_{n-q}(c) \tag{15}
$$

   a matrix with columns ranging from 0 to $n - (\deg b + \deg c)$.

4. *The convolution matrix of a vector with single column is the vector itself.*

$$
\mathfrak{A}_p(c) = c \tag{16}
$$

5. *A polynom coefficient vector d is element of* $\operatorname{Im} \mathfrak{A}_n(c)$ *if and only if the related polynom is divisable by the polynom of c.*

6. *If the polynom related to c divides the polynom of d, then the Moore-Penrose inverse of d*

$$
b = \left(\mathfrak{A}_n(c)^* \mathfrak{A}_n(c)\right)^{-1} \mathfrak{A}_n(c)^* d \tag{17}
$$

   defines the remaining factor b with $d = c * b$. This remark is useful, because the Euclidian algorithm used for the theoretical factorization is numerically not stable.

7. *The representation of a product by a Hankel matrix (8) with a polynom coefficient vector c*

$$G_{0:n-p,\ 0:p}\ c = \begin{bmatrix} g_{0+0} & g_{1+0} & \cdots & g_{p+0} \\ g_{0+1} & g_{1+1} & \cdots & g_{p+1} \\ \vdots & \vdots & & \vdots \\ g_{0+n-p} & g_{1+n-p} & \cdots & g_n \end{bmatrix} \begin{bmatrix} c_0 \\ c_1 \\ \vdots \\ c_p \end{bmatrix} = \begin{bmatrix} r_0 \\ r_1 \\ \vdots \\ r_{n-p} \end{bmatrix} \qquad (18)$$

can be written by reordering as

$$\begin{bmatrix} g_0 & g_1 & g_2 & \cdots & g_n \end{bmatrix} \mathfrak{A}_n(c) = \begin{bmatrix} r_0 & r_1 & \cdots & r_{n-p} \end{bmatrix} \qquad (19)$$

$$or \qquad\qquad G \qquad \mathfrak{A}_n(c) = \qquad R \qquad\qquad\qquad (20)$$

To see this, multiply the Hankel matrix row by row by c. This is the same as multiplying the whole set of measurements G with the shifted vector c as part of the matrix $\mathfrak{A}_n(c)$.

8. *Assume, that for a symmetric matrix H we have non negative constants v and μ such that*

$$v\ \mathfrak{A}(c)^*\mathfrak{A}(c) \leq \mathfrak{A}(c)^* H\ \mathfrak{A}(c) \leq \mu\ \mathfrak{A}(c)^*\mathfrak{A}(c) \qquad (21)$$

as operators, than for another coefficient vector b with an appropriate degree we have also

$$v\ \mathfrak{A}(b)^*\mathfrak{A}(c)^*\mathfrak{A}(c)\ \mathfrak{A}(b) \leq \mathfrak{A}(b)^*\mathfrak{A}(c)^* H\ \mathfrak{A}(c)\ \mathfrak{A}(b) \leq \mu\ \mathfrak{A}(b)^*\mathfrak{A}(c)^*\mathfrak{A}(c)\ \mathfrak{A}(b)$$

and

$$v\ \mathfrak{A}(b*c)^*\mathfrak{A}(b*c) \leq \mathfrak{A}(b*c)^* H\ \mathfrak{A}(b*c) \leq \mu\ \mathfrak{A}(b*c)^*\mathfrak{A}(b*c) \quad (22)$$

4 The Decomposition of a Signal in Koopman Modes

Using the roots λ_l of the polynom coefficient vector c with degree p for defining entities $v_l^f(q)$ in the next equation we will construct an **approximative** decomposition of $G^f(q)$ for all $q \in Q$ of the type

$$\mathbb{N}_0 \ni k \mapsto g_k^f(q) \approx \tilde{g}_k^f = \sum_{l=1}^{p} v_l^f(q)\ \lambda_l^{\ k} \quad \forall\ f \in S \qquad (23)$$

The p roots are related to the stability. $|\lambda_l| = 1$ describe unsteady but stable modes (typical); for $|\lambda_l| < 1$ the mode disappears; modes with $|\lambda_l| > 1$ are not stable. This describes the iterative development with respect to index k of all observables in S by **common** modes λ_l, the so-called Ritz values, which may be independent on q and f. The complex vectors $v_l(q) = \left(v_l^f(q) \right)_{f \in S}$ are named Koopman modes [4]. They are independent on the iteration number k. Even if the approximating sequence is not describing the complete signal, it gives the chance in getting insight in the behaviour of the underlying operator.

For $p = n$ the procedure is a reformulation of the so-called Dynamic Mode Decomposition of [7].

4.1 Vandermonde Decomposition for a Polynom Coefficient Vector

Let c with $\deg c = p$ be the coefficient vector of the polynom $c(\lambda) = \sum_{k=0}^{p} c_k \lambda^k \quad \forall \lambda \in \mathbb{C}$. We assume that **the p roots λ_l are pairwise different**. Otherwise the relations are more confusing; but we will not expect multiple roots of modulus 1 for stable cases. Each root λ_l, $l = 1, \ldots, p$ of the polynom c defines the coefficient-vector $w_l = \begin{bmatrix} w_{l0}, & w_{l1}, & \cdots, & w_{l\,p-1} \end{bmatrix}^T$ with $\deg w_l = p - 1$ and $w_l(\lambda_l) \neq 0$ by factorizing c

$$c(\lambda) = (\lambda - \lambda_l)\, w_l(\lambda) \quad \forall \lambda \in \mathbb{C} \quad \text{or} \quad c = w_l * \begin{bmatrix} -\lambda_l \\ 1 \end{bmatrix} \tag{24}$$

The p vectors w_l are linearly independent. To see this multiply $0 = \begin{bmatrix} 1, \lambda_m, \lambda_m^2, \ldots, \lambda_m^{p-1} \end{bmatrix} \sum_{l=1}^{p} \alpha_l\, w_l = \sum_{l=1}^{p} \alpha_l\, w_l(\lambda_m) = \alpha_m\, w_m(\lambda_m)$. Because λ_m is a simple root, $w_m(\lambda_m) \neq 0$, and we have $\alpha_m = 0$. It can be shown that (multiply by w_m from the right)

$$I_{0:p-1} = \sum_{l=1}^{p} \frac{1}{w_l(\lambda_l)} w_l \begin{bmatrix} 1, \lambda_l, \lambda_l^2, \ldots, \lambda_l^{p-1} \end{bmatrix} \tag{25}$$

is the identity operator. By this way every polynom coefficient vector c of degree p with pairwise different roots is associated with a p-dimensional subspace of polynom coefficient vectors. This will help to derive eigenmodes only from the set of roots or eigenvalues.

5 The Koopman Eigenvectors for a Decomposition

Rewriting equation (23) by stacking $(0 : p)$ subsequent elements leads to

$$\begin{bmatrix} \tilde{g}_k(q) \\ \tilde{g}_{k+1}(q) \\ \vdots \\ \tilde{g}_{k+p}(q) \end{bmatrix} = \sum_{l=1}^{p} \begin{bmatrix} v_l(q)\, \lambda_l^0 \\ v_l(q)\, \lambda_l^1 \\ \vdots \\ v_l(q)\, \lambda_l^p \end{bmatrix} \lambda_l^k = \sum_{l=1}^{p} v_l(q) \begin{bmatrix} \lambda_l^0 \\ \lambda_l^1 \\ \vdots \\ \lambda_l^p \end{bmatrix} \lambda_l^k \tag{26}$$

Multiplying from the left by a vector $u^* = \frac{w_i^*}{w_i(\lambda_i)}d^*$, where w_i is a polynom coefficient vector of degree $p - 1$ with $w_i(\lambda_l) = 0$ $\forall l \neq i$ and d_i is an arbitrary vector, this transforms the decomposition to the action on the single mode i

$$u_i^* \begin{bmatrix} \tilde{g}_k(q) \\ \tilde{g}_{k+1}(q) \\ \vdots \\ \tilde{g}_{k+p}(q) \end{bmatrix} = \sum_{l=1}^{p} d_i^* v_l(q) \frac{w_i(\lambda_l)}{w_i(\lambda_i)}\lambda_l^k = d_i^* v_i(q) \lambda_i^k \tag{27}$$

Returning back to the definition (5) of $\tilde{g}_k^f(q) \approx g_k^f(q) = f(\varphi^k q)$

$$u_i^* \begin{bmatrix} f(\varphi^k \circ \varphi\, q) \\ f(\varphi^{k+1} \circ \varphi\, q) \\ \vdots \\ f(\varphi^{k+p} \circ \varphi\, q) \end{bmatrix} \approx d_i^* v_i(q) \lambda_i^{k+1} \approx \lambda_i u_i^* \begin{bmatrix} f(\varphi^k q) \\ f(\varphi^{k+1}q) \\ \vdots \\ f(\varphi^{k+p}q) \end{bmatrix} \tag{28}$$

showing, that $u_i^* \left[f \circ \varphi^{k+j} \right]_{j=0,\cdots,p}$ approximates a Koopman eigenfunction for the eigenvalue λ_i on the trajectories starting with $q \in Q$. Remarkable is, that the approximate eigenfunction is composed by the values only on the specific trajectory belonging to $q \in Q$. Because d_i is an arbitrary vector, the eigenspace of λ_i is as large as the dimension of the linear space generated by the observables $f \in S$. This does not imply, that λ_i is a multiple root of c. For a case of having the vectors g_k as functions on a discrete set Ω, the vectors d_i may be evaluation operators, showing that the Koopman eigenfunctions can be understood themselves as discrete functions on the same set Ω.

5.1 Wiener-Wintner Eigenfunctions and Decomposition

Assumed is a measure μ on K with the property $\mu(\varphi^{-1}(A)) = \mu(A)$ for all measurable sets A of K. Then for μ-almost all $q \in K$ and for real ω the sum

$$\tilde{f}_\omega(q) = \lim_{N\to\infty} \frac{1}{N} \sum_{k=0}^{N-1} f(\varphi^k q) e^{i\,2\pi\,\omega k} \tag{29}$$

converges. This is the content of the theorem of Wiener-Wintners (1931), see [1]. Clearly we see that by

$$\tilde{f}_\omega(\varphi q) = e^{-i\,2\pi\omega}\tilde{f}_\omega(q) \quad \text{for } \mu\text{-almost all } q \in K \tag{30}$$

that $\left(e^{-i\,2\pi\omega},\,\tilde{f}_\omega\right)$ is an eigenpair. The sum remembers the inverse Discrete Fourier Transform but with arbitrary real ω instead of the rational fractions $\omega = \frac{m}{N}$. For the special case $\omega = 0$ the mean value along the trajectory is an eigenfunction with eigenvalue 1.

How does that fit the described context? For all $l = 0, 1, \cdots, N$, define angles by $\omega_l = \omega + \frac{l}{N}$ and eigenvalues $\lambda_l = e^{-i2\pi\omega_l}$. The convolution of the polynom coefficient vectors w_l and the linear factors $\begin{bmatrix} -\lambda_l \\ 1 \end{bmatrix}$ result in

$$
e^{-i2\pi(N-1)\omega_l}\, w_l * \begin{bmatrix} -\lambda_l \\ 1 \end{bmatrix} = e^{-i2\pi(N-1)\omega_l} \begin{bmatrix} 1 \\ e^{i2\pi\omega_l} \\ \vdots \\ e^{i2\pi(N-1)\omega_l} \end{bmatrix} * \begin{bmatrix} -\lambda_l \\ 1 \end{bmatrix} = \begin{bmatrix} -e^{i2\pi N\omega} \\ 0 \\ \vdots \\ 0 \\ 1 \end{bmatrix} = c_\omega
$$

(31)

with a coefficient vector c_ω and $w_l\,(\lambda_l) = N$ independent on l. For $l = 0$ we find the sequence generating the sum (29). That means, the Wiener-Wintner sequence is generated by c_ω belonging to the family of ω_l.

6   Polynom Coefficient Vector and Hankel Matrix

Theorem 6.1 *Assume a vector c with $\deg c = p$ for an approximating sequence $\tilde{G} = \begin{bmatrix} \tilde{g}_0 & \tilde{g}_1 & \dots & \tilde{g}_n \end{bmatrix}$, such that its product with the Hankel matrix **vanishes***

$$
0 = \tilde{G}_{0:n-p,\ 0:p}\ c = \begin{bmatrix} \tilde{g}_{0+0} & \tilde{g}_{1+0} & \cdots & \tilde{g}_{p+0} \\ \tilde{g}_{0+1} & \tilde{g}_{1+1} & \cdots & \tilde{g}_{p+1} \\ \vdots & \vdots & & \vdots \\ \tilde{g}_{0+n-p} & \tilde{g}_{1+n-p} & \cdots & \tilde{g}_n \end{bmatrix} \begin{bmatrix} c_0 \\ c_1 \\ \vdots \\ c_p \end{bmatrix}
$$

(32)

Than for the roots λ_l of c and the vectors w_l defined by $c = w_l * \begin{bmatrix} -\lambda_l \\ 1 \end{bmatrix} (l = 1, \cdots, p)$

$$
\begin{bmatrix} \tilde{g}_0 & \tilde{g}_1 & \dots & \tilde{g}_n \end{bmatrix} = \sum_{l=1}^{p} v_l \begin{bmatrix} 1, \lambda_l, \lambda_l^2, \dots, \lambda_l^n \end{bmatrix}
$$

(33)

with the vectors (w_l enlarged by 0 for components $p + 1, \cdots, n$)

$$v_l = \tilde{g}_{0:p-1} \frac{1}{w_l(\lambda_l)} \; w_l = \tilde{G} \frac{1}{w_l(\lambda_l)} \begin{bmatrix} w_l \\ 0 \\ \vdots \\ 0 \end{bmatrix} \tag{34}$$

A smaller p implies less eigenvalues λ_l used in the decomposition. The matrix $\tilde{G}_{0:n-p,\,0:p}$ must have more inner dependencies for smaller p.

Proof We have by (24) in (32) $\tilde{G}_{0:n-p,\,0:p} \; w_l * \begin{bmatrix} -\lambda_l \\ 1 \end{bmatrix} = \tilde{G}_{0:n-p,\,0:p} \; c = 0$ for every $l = 1, \cdots, p$
and therefore $\tilde{G}_{\underset{0:n-p,\,\mathbf{1:p}}{\downarrow}} w_l = \lambda_l \tilde{G}_{\underset{0:n-p,\,\mathbf{0:p-1}}{\downarrow}} w_l$ or row by row $\tilde{g}_{0+j+1:p+j} \; w_l = \lambda_l \tilde{g}_{0+j:p-1+j} \; w_l$ and by induction

$$\tilde{g}_{0+j+1:p+j} \; w_l = \lambda_l^{j+1} \tilde{g}_{0:p-1} \; w_l \quad \forall \, j = 0, \cdots, n-p \tag{35}$$

Using the Vandermonde decomposition (25) we get

$$\begin{bmatrix} \tilde{g}_0 & {}_{:p-1} \\ \tilde{g}_1 & {}_{:p} \\ & \vdots \\ \tilde{g}_{n-p} & {}_{:n-1} \\ \tilde{g}_{n-p+1} & {}_{:n} \end{bmatrix} = \sum_{l=1}^{p} \begin{bmatrix} \tilde{g}_0 & {}_{:p-1} \\ \tilde{g}_1 & {}_{:p} \\ & \vdots \\ \tilde{g}_{n-p} & {}_{:n-1} \\ \tilde{g}_{n-p+1} & {}_{:n} \end{bmatrix} \frac{w_l}{w_l(\lambda_l)} \left[1, \lambda_l, \lambda_l^2, \dots, \lambda_l^{p-1} \right]$$

$$\overset{(35)}{=} \sum_{l=1}^{p} \frac{1}{w_l(\lambda_l)} \tilde{g}_{0:p-1} \; w_l \begin{bmatrix} 1 \\ \lambda_l \\ \vdots \\ \lambda_l^{n-p} \\ \lambda_l^{n-p+1} \end{bmatrix} \left[1, \lambda_l, \lambda_l^2, \dots, \lambda_l^{p-1} \right] \tag{36}$$

by definition of the Hankel matrix (32) that means

$$\tilde{g}_k = \sum_{l=1}^{p} \frac{1}{w_l(\lambda_l)} \tilde{g}_{0:p-1} \; w_l \, \lambda_l^k \quad \forall \, k = 0, \cdots, n \tag{37}$$

such that we get finally the decomposition

$$\begin{bmatrix} \tilde{g}_0 & \tilde{g}_1 & \dots & \tilde{g}_n \end{bmatrix} = \sum_{l=1}^{p} v_l \left[1, \lambda_l, \lambda_l^2, \dots, \lambda_l^n \right] \tag{38}$$

6.1   Approximatve Decomposition of a Signal in Modes

Given is an arbitrary coefficient vector c with deg $c = p$. We will described who G can be decomposed in two parts $G = G_{modes} + \Delta G$ related to c. The first part G_{modes} will be given by a linear decomposition in modes defined by the roots of c and by the requirement $G_{modes}\, \mathfrak{A}(c) = 0$ and will described later. The second part defines the defect in relation (20) given by $\Delta G\, \mathfrak{A}(c) = R$. There are some reasonable requirements in selecting ΔG. First we restrict ΔG by

$$\text{Im } \Delta G \subset \text{Im } R \tag{39}$$

that means $\Delta G = R\, \beta^*$ for some matrix β and consequentially $R\, \beta^* \mathfrak{A}(c) = R$. For a general non restricted R, this enforces $\beta^* \mathfrak{A}(c) = I$. Defining $\beta^* = (\alpha^* \mathfrak{A}(c))^{-1} \alpha^*$ fulfills this condition, if $\alpha^* \mathfrak{A}(c)$ is invertible. We simply take $\alpha = \mathfrak{A}(c)$, but this not mandatory. With these assumptions we get (remember, that $\mathfrak{A}(c)$ has full rank)

$$R\, \mathfrak{A}(c)^* = R\, \left(\mathfrak{A}(c)^* \mathfrak{A}(c)\right)^{-1} \mathfrak{A}(c)^* \overset{(20)}{=} G\, \mathfrak{A}(c)\, \left(\mathfrak{A}(c)^* \mathfrak{A}(c)\right)^{-1} \mathfrak{A}(c)^* = G\, Q$$

and therefore

$$\Delta G = R\, \mathfrak{A}(c)^* = G\, Q \tag{40}$$

for the selfadjungated projection Q $(Q = Q^2,\ Q = Q^*)$ on Im ΔG

$$Q = \mathfrak{A}(c)\, \left(\mathfrak{A}(c)^* \mathfrak{A}(c)\right)^{-1} \mathfrak{A}(c)^* \tag{41}$$

with the property

$$(I - Q)\, \mathfrak{A}(c) = 0 \tag{42}$$

This implies for the decomposition $G = G_{modes} + \Delta G = G\,(I - Q) + G\,Q$ and the property $0 = G_{modes}\, \mathfrak{A}(c) = \begin{bmatrix} \tilde{g}_0 & \tilde{g}_1 & \dots & \tilde{g}_n \end{bmatrix} \mathfrak{A}(c)$. Switching to the equivalent Hankel representation of $G_{modes} = \tilde{G}$ we have (32) and the decomposition (33) and therefore

$$G_{modes} = \begin{bmatrix} \tilde{g}_0 & \tilde{g}_1 & \dots & \tilde{g}_n \end{bmatrix} = \sum_{l=1}^{\mathbf{p}} v_l \left[1, \lambda_l, \lambda_l^2, \dots, \lambda_l^{\mathbf{n}} \right] \tag{43}$$

with the modes $v_l = G_{modes}\, \frac{1}{w_l(\lambda_l)} \begin{bmatrix} w_l \\ 0 \end{bmatrix}$ from (34).

We pave now the way for quantification of the l_2-norm $\| \Delta G \|_2$ of the defect operator ΔG. Using the l_2-norm is a clear restriction. It might be more appropiate to investigate the l_∞-norm. Taking the quantity $\mu = \| \Delta G \|_2^2$ we have to analyse the operator inequality $\Delta G^* \Delta G \leq \mu I$ or by (40)

$$Q^* H Q \leq \mu I \tag{44}$$

with the SPD matrix $H = G^T G$ and the projection Q in (41).
The operator inequality (44) with explicit Q can be written as

$$\mathfrak{A}(c) \left(\mathfrak{A}(c)^* \mathfrak{A}(c)\right)^{-1} \mathfrak{A}(c)^* \; H \; \mathfrak{A}(c) \left(\mathfrak{A}(c)^* \mathfrak{A}(c)\right)^{-1} \mathfrak{A}(c)^* \leq \mu I. \tag{45}$$

Because $\mathfrak{A}(c)$ has full rank, this is equivalent to

$$\mathfrak{A}(c)^* \; H \; \mathfrak{A}(c) \leq \mu \, \mathfrak{A}(c)^* \, \mathfrak{A}(c) \tag{46}$$

We end up with the following theorem.

Theorem 6.2 *Given is an arbitrary coefficient vector c with* deg $c = p$. *Assume that the polynom c has no multiple roots.*
We can decompose G in two parts

$$G = G_{modes} + \Delta G \tag{47}$$

where for $\Delta G = G \, Q$ with $Q = \mathfrak{A}(c) \left(\mathfrak{A}(c)^* \mathfrak{A}(c)\right)^{-1} \mathfrak{A}(c)^*$ fulfilling the requirements (39)
we have $\|\Delta G\|_2 \leq \sqrt{\mu}$ iff

$$\mathfrak{A}(c)^* \; H \; \mathfrak{A}(c) \leq \mu \, \mathfrak{A}(c)^* \, \mathfrak{A}(c) \tag{48}$$

For the roots λ_l of c and $v_l = \frac{1}{w_l(\lambda_l)} G_{modes} \begin{bmatrix} w_l \\ 0 \end{bmatrix}$ from (33) with $c = w_l * \begin{bmatrix} -\lambda_l \\ 1 \end{bmatrix}$

$$G_{modes} = \sum_{l=1}^{p} v_l \left[1, \lambda_l, \lambda_l^2, \ldots, \lambda_l^n\right] \tag{49}$$

6.2 Simplified Approach

Applying the **trace** on both sides of operator inequality (48), we get by definition of $H = G^T G$ for the j-th shifted row $G^j = \begin{bmatrix} g_{0+j} & g_{1+j} & \cdots & g_{n+j} \end{bmatrix}$

$$\frac{1}{n-p+1} \sum_{j=0}^{n-p} \left\|G^j \, c\right\|^2 \leq \frac{\mu}{n-p+1} \sum_{j=0}^{n-p} \|c\|^2 = \mu \, \|c\|^2 \tag{50}$$

or with the collapsed matrix H^{n-p} which is composed by a sum of shifted submatrices of H

$$H^{n-p} = \frac{1}{n-p+1} \sum_{j=0}^{n-p} \left(G^j\right)^T G^j \tag{51}$$

we have for the pair (c, μ)

$$< H^{n-p} c, c > \leq \mu \parallel c \parallel^2 \tag{52}$$

suggesting to take c as eigenvector of H^{n-p} with the smallest eigenvalue. But inequality is only a necessary and not sufficient condition for (48).

7 Remarks

1. With this operator valued estimate it is possible to calculate the approximation quality of the modes defined by the roots of c for any polynom coefficient vector c with degree p.
2. We have an algorithm approximating the minimal value μ and the respective c for a given degree p. It uses Rellichs theorem [10] that an eigenvalue problem with symmetric matrices depending analytically on a real parameter provides eigenvalues and eigenvectors also depending analytically on that parameter.
3. The algorithm delivers eigenvalues with modulus near to 1 for our test cases. Some of these fulfill the condition $\lambda_j \lambda_k \approx \lambda_{m(j,k)}$ for some j, k. The calculation of the eigenvalues is stiff. Some small changes in c may change some of the eigenvalues significantly. We will try to control this by the later condition.
4. Changing the degree p has influence on some eigenvalues but not on all.
5. The degree p of c should be small to limit the number of modes; on the other hand a small degree enlarges the approximation error μ. For the largest possible $p = n$, we have $\mathfrak{A}(c) = c$ and (v, μ) can be an eigenpair for smallest eigenvalue of H. This is the setting for the Dynamic Mode Decomposition (DMD) of [7].
6. The simplified approach might help to find appropriate polynom coefficient vectors c. The related constant μ is underestimating the true error. But if μ is small and $n - p$ is not large, then c might be appropriate.
7. Whereas G might be a very large matrix with many rows, H is a quadratic matrix having the number of e.g. time steps as dimension which is typically smaller.
8. If (48) is given for several G respective H coming from different test cases for a common small μ, then all these test cases share a common decomposition with the same Ritz-values λ_l. In this way ensembles could be handled.

8 Conclusions

Extending the work in [9] we try to understand how to transfer the idea and the properties of the Koopman operator from functional analysis to stimulate the analysis of non wellposed simulation problems.

We show how to determine eigenvalues as roots of a vector, which is near to the kernel of the matrix of measurements in a specific sense. The roots enable the construction of the so-called Koopman modes as well the Koopman eigenfunctions. The resulting roots are only partially near to Koopman eigenvalues. To identify the spurious eigenvalues and to delete these or to shift these to a reasonable position is important. We think, that the property of the Koopman operator, that products of eigenvalues are again eigenvalues (if the product of their eigenfunctions does not disappear), might help to identify true Koopman eigenvalues and eigenfunctions.

References

1. Eisner, T., Farkas, B., Haase, M., Nagel, R.: Operator Theoretic Aspects of Ergodic Theory. Graduate Texts in Mathematics. Springer (2015)
2. Koopman, B.O.: Hamiltonian systems and transformations in Hilbert space. Proc. Natl. Acad. Sci. USA **17**(5), 315318 (1931)
3. Mezić, I.: Spectral properties of dynamical systems, model reduction, and decompositions. Nonlinear Dyn. **41**(1–3), 309325 (2005)
4. Budišić, M., Mohr, R., Mezić, I.: Applied Koopmanism. Chaos 22, 047510 (2012). doi:10.1063/1.4772195
5. Chen, K.K., Tu, J.H., Rowley, C.W.: Variants of dynamic mode decomposition: boundary condition, Koopman, and Fourier analyse. J. Nonlinear Sci. **22**(6), 887915 (2012)
6. Peller, V.V.: An excursion into the theory of Hankel operators, in Holomorphic spaces (Berkeley, CA, 1995). In: Math. Sci. Res. Inst. Publ., Cambridge Univ. Press, Cambridge, vol. 33, pp. 65120 (1998)
7. Schmid, P.J.: Dynamic mode decomposition of numerical and experimental data. J. Fluid Mech. **656**, 24 (2010)
8. Rowley, C.W., Mezić, I., Bagheri, S., Schlatter, P., Henningson, D.S.: Spectral analysis of nonlinear flows. J. Fluid Mech. Cambridge University Press (2009)
9. Küster, U.: The spectral structure of a nonlinear operator and its approximation. In: Sustained Simulation Performance 2015: Proceedings of the joint Workshop on Sustained Simulation Performance, University of Stuttgart (HLRS) and Tohoku University, 2015, pp. 109–123, Springer International Publishing, ISBN:978-3-319-20340-9, doi:10.1007/978-3-319-20340-9_9
10. Rellich, F.: Störungstheorie der Spektralzerlegung I., Analytische Störung der isolierten Punkteigenwerte eines beschränkten Operators. Math. Ann. **113**, 600–619 (1937)

Implementation of a Parallel Sparse Direct Solver on Vector Architecture

Atsushi Suzuki and François-Xavier Roux

Abstract Linear systems with large sparse matrices are solved in finite element analysis of elasticity and/or fluid problems. Thanks to development of graph partitioning software, it becomes feasible to extract dense sub-matrices efficiently with minimizing fill-in during factorization. By analyzing task dependency of block factorization of dense matrix, multi-cores of CPUs which share the main memory are used in parallel and asynchronously. The tasks in dense sub-matrices consist of BLAS level 3 kernels which efficiently use arithmetic capabilities of modern super-scalar CPU with large cache memory and also of modern vector CPU. BLAS level 3 kernels can also efficiently use vector architecture, without writing any directives for explicit vectorization in the code. Nevertheless, the sparse part still remains in factorization process. Although it is only a small fraction of the whole process and almost negligible on the super-scalar CPU, its optimization is important on vector architecture due to short vector loop.

1 Introduction

We deal with large sparse matrices obtained from discretization by finite element methods, and we assume that unsymmetric $N \times N$ matrix A has symmetric structure of non-zero entries and has an LDU-factorization with symmetric partial pivoting,

$$A = \Pi^T L D U \Pi . \qquad (1)$$

Here L and U are a unit lower or upper triangle, respectively. When the matrix has k-dimensional null space, the last k entries of the diagonal matrix D become zero.

A. Suzuki (✉)
Cybermedia Center, Osaka University, Machikaneyama,
Toyonaka, Osaka 560-0043, Japan
e-mail: atsushi.suzuki@cas.cmc.osaka-u.ac.jp

F.-X. Roux
LJLL, UPMC (Paris 6)/ONERA, 4 place Jussieu, 75005 Paris, France
e-mail: roux@ann.jussieu.fr

© Springer International Publishing AG 2016
M.M. Resch et al. (eds.), *Sustained Simulation Performance 2016*,
DOI 10.1007/978-3-319-46735-1_8

We apply standard nested-dissection [1] algorithm to obtain a multi-frontal factorization strategy for effective parallel computation. Since nested-dissection is based on recursive bisection procedure of graph partitioning, sparse sub-matrices lie in the lowest level of the bisection tree, and number of sub-matrices in each level is two to the power of the level number. Parallel execution of factorizations of sparse sub-matrix in the lowest level is natural even though variation of size of sub-matrices needs to be taken account. It is mandatory to parallelize dense matrix in the higher level of the bisection tree, in multi-core CPU environment. By introducing block strategy for factorization of dense matrix, both utilization of BLAS level 3 kernels suitable to cache memory architecture and parallelization become possible. However, this block strategy changes pivoting procedure drastically because search range of diagonal entries during pivoting is limited within the block size, which may request usage of combination of 1×1 and 2×2 pivoting entries. Hence, we need to consider factorization with symmetric permutation,

$$A = \Pi^T \tilde{L} \tilde{D} \tilde{U} \Pi \tag{2}$$

where $\tilde{D}$ consists of mixture of 1×1 and 2×2 blocks. In practice, usage of 2×2 pivoting is minimized by recomputing Schur complement from postponed entries. Here diagonal entry is postponed when the ratio to a successive entry becomes less than a given threshold. The postponing strategy is rather common technique and dynamic data management is used for implementation [2]. In our strategy, null pivot candidates are postponed in the last Schur complement, which brings usage of static data management.

Many of modern supercomputers consist of cluster of multi-core CPU system, super-scalar core combined with large sized cache memory or vector core with small sized cache memory. On the both systems, BLAS level 3 kernels efficiently use arithmetic units of the core, which can hide difference of architecture, super-scalar or vector arithmetic units, and the same code can run on the both systems.

We aim to run the code on shared memory architecture, where data transfer between CPUs is no necessary. For distributed environment, LDU-factorization of sparse matrix is used as a subdomain solver of domain decomposition algorithms, for example, FETI [3] and BDD [4] methods.

In Sect. 2, factorization strategies using block substructure in both sparse and dense sub-matrices are discussed. In sparse matrix computation, implementation to vector architecture takes different approach than to super-scalar architecture for computational efficiency. Asynchronous parallel execution of block elimination of dense matrix is reviewed from previous paper [5]. In Sect. 3, difference of computational efficiency between super-scalar and vector architectures is reported by using a finite element matrix from incompressible Navier-Stokes equations. In the last section, we conclude our results.

2 Factorization of Sparse and Dense Sub-matrices

2.1 Recursive Computation of Schur Complement

By bisection procedure, the whole sparse matrix is decomposed into a union of an interface that produces a dense sub-matrix and two sparse sub-matrices. This procedure is applied recursively and a decomposition into mixture of sparse and dense sub-matrices is obtained. Fig. 1 shows an example of three-level nested-dissection by analogy with decomposition of a two dimensional domain and corresponding binary tree. To get such a nested-dissection graph partitioning, software packages METIS [6] and SCOTCH [7] which reduce fill-in and optimize number of multi-fronts for parallel computation are available.

In each level of the bisection tree, sub-matrix is factorized as

$$
\begin{bmatrix} A_{11} & A_{12} \\ A_{21} & A_{22} \end{bmatrix} = \begin{bmatrix} L_{11} & 0 \\ A_{21}U_{11}^{-1}D_{11}^{-1} & I_2 \end{bmatrix} \begin{bmatrix} D_{11} & 0 \\ 0 & S_{22} \end{bmatrix} \begin{bmatrix} U_{11} & D_{11}^{-1}L_{11}^{-1}A_{12} \\ 0 & I_2 \end{bmatrix},
\tag{3}
$$

with $S_{22} = A_{22} - A_{21}A_{11}^{-1}A_{12} = A_{22} - A_{21}(L_{11}D_{11}U_{11})^{-1}A_{12}$. Here S_{22} is computed from data of one side of domains sharing the interface and is gathered with the other side for the next level of the bisection tree.

In the lowest level of the bisection tree, sparse matrix A_{11} is factorized with tridiagonal block structure explained in the following subsection. We note that A_{12} consists of sparse matrix and needs to be dealt by using sparse data structure. For unsymmetric matrix, $A_{21}U_{11}^{-1}$ is computed by $U_{11}^{-T}A_{21}^{T}$, where data are stored in a transposed way with the same structure as for A_{12}. This treatment is optimal because structural symmetry on non-zero entries is assumed.

In other levels of the bisection tree, A_{11} is dense matrix and A_{12} and A_{21} consist of strip structure inheriting sparsity of the original matrix. The dense sub-matrix A_{11} is factorized by introducing block strategy.

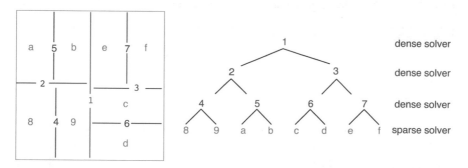

Fig. 1 Schematic figure of nested-dissection of two dimensional domain (*left*) and corresponding binary tree (*right*)

In the following, we discuss two kinds of block structure for factorization to reduce computational complexity for the sparse layer, and to introduce parallel computation for the dense layer.

2.2   Tridiagonal Block Structure for Sparse Sub-matrix

The sparse matrix is renumbered into a block tridiagonal structure with variable block size by reverse Cuthill-McKee ordering [8], which is similar to a uni-frontal approach. Let N_s be the size of the square sparse matrix. Left side of Fig. 2 shows tridiagonal block structure with variable block size b_m. The first diagonal block is factorized into LDU denoted as task $\alpha_1^{(1)}$, forward substitutions of the linear system with the sparse matrix for multiple right-hand sides are performed for both upper ($\beta_{+2}^{(1)}$) and lower blocks ($\beta_{-2}^{(1)}$), and then rank-b_2 update ($\gamma_2^{(1)}$) is performed to generate the second diagonal block. Here factorization procedure is completely serial. However, number of node in the lowest bisection level is usually large and then there are a lot of independent tasks for sparse factorization. In order to reduce complexity of computation of Schur complement, multiple right hand sides (RHS) which are denoted as A_{12} or A_{21}^T are sorted in column by decreasing order of the height of column vector which is defined by the first non-zero entry in the row. This is depicted in right side of Fig. 2. By using this reordering, sparse RHS is formed in block structure and forward substitution of some blocks can be completely or partially omitted due to zero entries, which achieves reduction of complexity in computation of $L_{11}^{-1} A_{12}$.

Since the size of sparse matrix is usually given as almost same order of the length of the vector register, 256, block size b_m is too small for vectorization. Therefore, on the vector architecture, after completion of LDU factorization of the tridiagonal matrix, the factorized matrix is stored in $N_s \times N_s$ dense matrix and the standard DTRSM routine for triangular solver in BLAS is called for forward substitution of the lower triangle and transposed forward substitution without reordering of multiple RHS. Efficiency of this implementation is reported in Sect. 3.

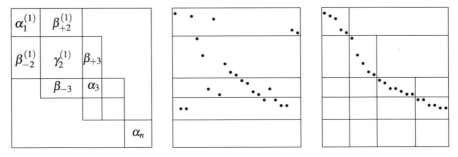

Fig. 2 Tridiagonal blocks sparse matrix (*left*), forward substitution of condensed sparse multiple right hand side (*center*), and sorted multiple RHS, where the first nonzero entry in each row is shown (*right*)

2.3 Block Strategy to Enforce Parallelization for Dense Factorization

The dense matrix is decomposed into a block structure with constant block size b. Right side of Fig. 1 shows the first stage of block elimination with three tasks, α: LDU-factorization, β_+, β_- : forward substitution and scaling without/with transposing realized by DTRSM, γ : rank-b update realized by DGEMM. On the super-scalar architecture this block size b is taken as large as possible so that the blocks can fit in cache memory to archive highest possible ratio between number of arithmetic operations and number of floating point data access. In the following numerical experiment described in Sect. 3, b is set as 480.

During factorization, when an entry of the diagonal block becomes smaller than the previous value with a certain ratio, the factorization is terminated and treatment for the rest of entries is postponed. At the end of factorization procedure, Schur complement of postponed null pivots is examined by a new kernel (null space) detection algorithm [5]. This postponing procedure is implemented using static data structure and overall parallel efficiency is not deteriorated in usual case when the matrix has a small dimensional null space or is invertible.

A task queue for block elimination of dense matrix following the first, second, third stages depicted in Fig. 3 and so on is made as

$$
\begin{aligned}
\alpha_1^{(1)} &\leftarrow \{\beta_{+2}^{(1)}\text{-}\beta_{-2}^{(1)}\text{-}\gamma_{2,2}^{(1)}\text{-}\alpha_2^{(2)}, \beta_{+3}^{(1)}, \beta_{-3}^{(1)}\beta_{+4}^{(1)}, \beta_{-4}^{(1)}, \ldots, \beta_{+n}^{(1)}, \beta_{-n}^{(1)}\} \\
&\leftarrow \{\gamma_{2,3}^{(1)}, \gamma_{3,3}^{(1)}, \ldots, \gamma_{3,2}^{(1)}, \ldots, \gamma_{n,n}^{(1)}\} \\
&\leftarrow \{\beta_{+3}^{(2)}\text{-}\beta_{-3}^{(2)}\text{-}\gamma_{3,3}^{(2)}\text{-}\alpha_3^{(3)}, \beta_{+4}^{(2)}, \beta_{-4}^{(2)}, \ldots, \beta_{+n}^{(2)}, \beta_{-n}^{(2)}\} \\
&\leftarrow \{\gamma_{3,4}^{(2)}, \ldots, \gamma_{4,3}^{(2)}, \ldots, \gamma_{n,n}^{(2)}\} \leftarrow \cdots \\
&\leftarrow \beta_{+n}^{(n-1)}\text{-}\beta_{-n}^{(n-1)}\text{-}\gamma_{n,n}^{(n-1)}\text{-}\alpha_n^{(n)} .
\end{aligned}
\tag{4}
$$

Here the symbol '$\leftarrow$' shows a dependence between tasks. Tasks in braces '{' and '}' do not depend on each other. Four ones connected with the symbol '-' show

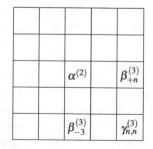

Fig. 3 Tasks of block elimination for dense matrix: 1st procedure (*right*), 2nd (*center*) and 3rd (*left*)

sequentially executed ones in a single core. This definition of ones dependency allows asynchronous parallel execution. In the computational code, tasks are assigned to cores of CPU by POSIX threads [9]. Detail of task management strategy is found in [5].

3 Efficiency of the Solver on Both Super-Scalar and Vector Architecture

"Dissection" code is written by C++, compiled by Intel C++ compiler ver. 16.0.2 and linked with SCOTCH ver. 6.0.4 [10] and sequential BLAS library in Intel MKL ver. 11.3.2 [11], or compiled by C++/SX rev. 103 and linked with sequential BLAS library in MathKeisan provided by NEC.

Numerical performance of Dissection is measured for a finite element matrix to solve stationary three dimensional Navier-Stokes equations with Re=300 in a cubic cavity box whose dimensionless size is 1. Finite element pair P2/P1 is used to approximate velocity and pressure, respectively. The matrix is unsymmetric and it has one dimensional null space due to pressure underdetermination. Mesh size of finite element tetrahedron is $1/34$, by excluding degrees of freedom given as Dirichlet data, number of DOF is 945,164, and number of nonzero entries of the sparse matrix is 89,588,848.

Two multi-core systems with different architecture, two noded super-scalar CPU and one vector CPU are used : Intel Xeon E5-2695v3 running at 2.3GHz with peak performance 36.8GFlop/s by one core and 1,030.4GFlop/s by 28 cores, and NEC SX-ACE at 1GHz, 64.0GFlop/s by one core and 256.0GFlop/s by four cores. Both system have 64GB main memory.

As a comparison of parallel efficiency on super-scalar architecture, Intel Pardiso in MKL ver. 11.3.2 [11] is used on Intel Xeon. Since the matrix is singular and the symmetric part of the matrix is not positive definite, adequate pivoting strategy and capability to deal with rank deficiency are mandatory. Intel Pardiso uses $\sqrt{\varepsilon}$-perturbation strategy combined with pivoting during factorization [12], and as a result it has no capability to detect the null space of the matrix. Obtained error of the solution is 5.7133×10^{-15} by Dissection and 5.2005×10^{-3} by Intel Pardiso, respectively. Table 1 shows parallel efficiency of Dissection and Pardiso solvers with elapsed time in seconds and CPU time. Dissection on Intel Xeon has good strong scalability about 48 % for 28 cores. Increase of CPU time in Dissection is smaller than in Pardiso, which coincides with parallel efficiency. This is caused by usage of sequential BLAS library and asynchronous parallelization where coarse grain parallelization is realized.

The top of Fig. 4 shows timelines on Intel Xeon with 1, 2 and 4 cores, where we can observe well assigned parallel tasks that are asynchronously executed. Distinguishing five colors correspond to major tasks; purple to the sparse factorization with tridiagonal data structure explained in Sect. 2.2, yellow to computing local

Table 1 Parallel efficiency of finite element matrix from Navier-Stokes equations ($n = 945,164$, $nnz = 89,588,848$), CPU and elapsed time in seconds

# of cores	Intel Pardiso (Intel Xeon)			Dissection (Intel Xeon)			Dissection (NEC SX-ACE)		
	CPU	Elapsed	Speed-up	CPU	Elapsed	Speed-up	CPU	Elapsed	Speed-up
1	2,052.4	2,053.3	—	1,268.0	1,268.9	—	1,080.4	1081.9	—
2	2,479.3	1,255.7	×1.64	1,269.9	659.39	×1.95	1,108.3	590.96	×1.83
4	2,757.6	698.19	×2.94	1,350.6	356.22	×3.56	1,178.5	345.84	×3.12
8	3,535.8	448.13	×4.58	1,469.2	201.24	×6.31			
16	4,556.9	298.88	×6.87	1,813.2	129.63	×9.78			
24	5,246.1	222.33	×9.24	1,879.8	96.18	×13.19			
28	5,322.4	191.76	×10.71	2,002.0	94.34	×13.45			

Schur complement matrix of the sparse layer, dark blue to DGEMM operation within LDU-factorization of the dense layer which is explained as task γ in Sect. 2.3, light blue to DGEMM operation to compute Schur complement matrix of the dense layer, light green to DTRSM operation for forward substitution of multiple RHS in the off-diagonal block, i.e., computation of $L_{11}^{-1}A_{12}$ and $U_{11}^{-T}A_{21}^{T}$ in the dense layer.

The center of Fig. 4 shows timelines of Dissection code on SX-ACE without modification, where BLAS routines are linked to ones in MathKeisan and no explicit vector directive is added. We observe that elapsed time for DGEMM shown by light blue, DTRSM by light green in the dense layer are reduced than those on Intel Xeon. However, elapsed time for the sparse layer takes around four times, which is clearly seen by one core execution. Yellow color is dominant in one core execution, which indicates that computation of local Schur complement including permutation of multiple RHS, forward substitution with tridiagonal structure is not efficiently performed by vector units because of short loops.

The bottom of Fig. 4 shows timelines of Dissection code on SX-ACE with modification to use DTRSM for the whole forward substitution in the sparse matrix and no ordering for multiple RHS, discussed in the end of Sect. 2.2. Performance of the sparse layer is improved and elapsed time becomes less than half by this strategy.

We note that purple color becomes dominant with four cores and sum of elapsed time with four cores is bigger than time with one core. This indicates that sparse factorization with tridiagonal structure is memory bounded and four cores seem to cause conflicts for accessing the main memory.

Table 2 shows measured GFlop/s of DGEMM kernel in the dense layer on both Intel Xeon E5-2695v3 and NEC SX-ACE. DGEMM kernal achieves 65 % of the peak performance on SX-ACE by calling sequential one and 62 % on Intel Xeon with 28 cores that run at 2.3 GHz. Since Intel Xeon E5-2695v3 has capability to increase running frequency up to 3.3 GHz, average performance of DGEMM by one core is almost same as the theoretical peak performance of 2.3 GHz.

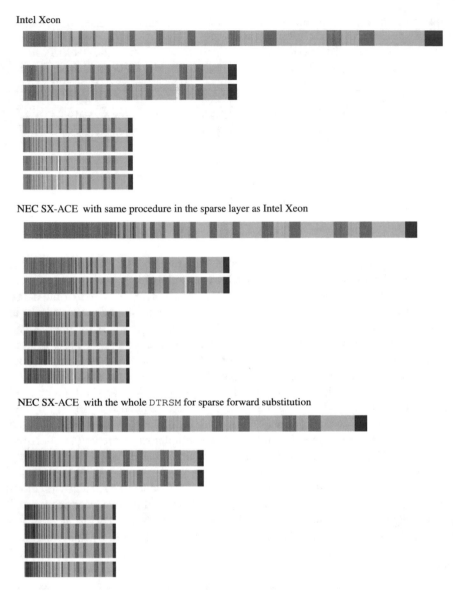

Fig. 4 Timelines of Dissection solver on Intel Xeon (*top*) NEC SX-ACE (*center and bottom*) with 1, 2 and 4 cores

Though the exact same source code written by C++ is used, there are difference in memory usage; 46.62 GB for Intel Xeon and 59.94 GB for SX-ACE, due to some difference of dynamical memory management by the operating system.

Table 2 GFlop/s of DGEMM BLAS level 3 kernel in the dense layer of Dissection

# of cores	Intel Xeon	NEC SX-ACE
1	36.35	44.85
2	36.27	43.76
4	34.06	41.31
8	31.23	
16	25.25	
24	24.38	
28	22.90	

4 Conclusion

Parallel sparse direct solver named as "Dissection" is developed for multi-core CPU with shared memory architecture. Nested-dissection graph partitioning provides mixture of factorization in the sparse layer and the dense layer. The code is written by C++ and aimed to run on super-scalar CPU efficiently, because the major operation of the code consists of BLAS level 3 operations. Dependency analysis of tasks for block factorization achieves asynchronous parallel execution realized by POSIX threads. Parallel efficiency better than Intel Pardiso on Intel Xeon CPU is verified by a numerical example using a finite element matrix from incompressible Navier-Stokes equations.

On NEC SX-ACE, which has modern vector arithmetic architecture, it is not necessary to introduce explicit vector directives in the code, and optimization for vectorization is carried out by calling vectorized sequential BLAS level 3 kernels. However some inefficiency remains for vector units in the sparse layer of the code, which consumes only 6 % of the elapsed time in super-scalar architecture. This part becomes a bottleneck because BLAS level 3 computation in the dense layer on vector CPU becomes faster than on super-scalar CPU and the block size of sparse part is not enough for vectorization due to sparse structure. One idea to improve vector efficiency is shown. It consists in replacing small block structure by large one thanks to ignoring sparse structure after factorization completed.

We conclude that the usage of sequential BLAS level 3 kernels is not only efficient on super-scalar architecture but also on modern vector CPU, which allows us to have a common code that is independent of the architecture, and asynchronous parallel execution well utilizes multi-core CPUs.

Acknowledgements This work is partially supported by "Joint Usage/Research Center for Interdisciplinary Large-scale Information Infrastructures" in Japan. Computational time for Cray XC30 in Institute for Information Management and Communication, Kyoto University, and for NEC SX-ACE in Cybermedia Center, Osaka University, are provided by this grant.

References

1. George, A.: Numerical experiments using dissection methods to solve n by n grid problems. SIAM J. Numer. Anal. **14**, 161–179 (1977). doi:10.1137/0714011
2. Amestoy, P.R., Duff, I.S., L'Excellent, J.-Y.: Multifrontal parallel distributed symmetric and unsymmetirc solvers. Comput. Meth. Appl. Mech. Eng. **184**, 501–520 (2000). doi:10.1016/S0045-7825(99)00242-X
3. Farhat, C., Roux, F.-X.: Implicit parallel processing in structural mechanics. Comput. Mech. Adv. **2**, 1–124 (1994)
4. Mandel, J.: Balancing domain decomposition. Commun. Numer. Meth. Eng. **9**, 233–241 (1993). doi:10.1002/cnm.1640090307
5. Suzuki, A., Roux, F.-X.: A dissection solver with kernel detection for symmteric finite element martices on shared memory computers. Int. J. Numer. Meth. Eng. **100**, 136–164 (2014). doi:10.1002/nme.4729
6. Karypis, G., Kumar, V.: A fast and high quality multilevel scheme for partitioning irregular graphs. SIAM J. Sci. Comput. **20**, 359–392 (1998). doi:10.1137/S1064827595287997
7. Pellegrini, F., Roman, J., Amestoy, P. : Hybridizing nested dissection and halo approximate minimum degree for efficient sparse matrix ordering. Concurr.: Pract. Experience **12**, 69–84 (2000)
8. George, A., Liu, J.W.H.: Algorithms for matrix partitioning and the numerical solution of finite element systems. SIAM J. Numer. Anal. **15**, 297–327 (1978). doi:10.1137/0715021
9. Lewis, B., Berg, D.J.: Multithreaded Programming with Pthreads. Sun Microsystems Press (1998)
10. Web site of Soctch and PT-Scotch. https://www.labri.fr/perso/pelegrin/scotch. Accessed 9 Sep 2016
11. Web site of Intel Math Kernel Library. http://software.intel.com/en-us/intel-mkl. Accessed 9 Sep
12. Schenk, O., Gärtner, K.: On fast factorization pivoting methods for sparse symmetric indefinite systems. Electron. Trans. Numer. Anal. **23**, 158–179 (2006)

Directive Translation for Various HPC Systems Using the Xevolver Framework

Kazuhiko Komatsu, Ryusuke Egawa, Hiroyuki Takizawa and Hiroaki Kobayashi

Abstract This paper proposes a directive translation approach that translates a special placeholder to different directives, depending on the target HPC system. The special placeholder in an application code is used as a trigger for the directive translation. By employing a code translation framework, *Xevolver*, the special placeholder can be translated to various directives that fit to any target HPC systems. Instead of using multiple directives, it can keep maintainability and readability of the original code because only special placeholders are inserted into an application code. This paper also demonstrates a translation of a special placeholder into OpenMP directives to clarify the effectiveness of the proposed directive translation approach.

1 Introduction

A huge number of transistors can be utilized to design a processor by the advancements in semiconductor technologies. As this has brought a wide design space of processors, various processors have been developed. For example, there are a scalar processor of multiple cores and large cache memories, an accelerator of a lot of small simple computational cores, and a vector processor of several cores specialized for vector calculations with a high memory bandwidth.

K. Komatsu (✉) · R. Egawa
Cyberscience Center, Tohoku University, 6-3 Aramaki-aza-aoba, Aoba,
Sendai 980-8578, Japan
e-mail: komatsu@tohoku.ac.jp

R. Egawa
e-mail: egawa@tohoku.ac.jp

H. Takizawa · H. Kobayashi
Graduate School of Information Sciences Tohoku University,
6-6-01 Aramaki-aza-aoba, Aoba, Sendai 980-8579, Japan
e-mail: takizawa@tohoku.ac.jp

H. Kobayashi
e-mail: koba@tohoku.ac.jp

© Springer International Publishing AG 2016
M.M. Resch et al. (eds.), *Sustained Simulation Performance 2016*,
DOI 10.1007/978-3-319-46735-1_9

109

In addition, the diversity of HPC systems has naturally increased by the various kinds of processors. Recent statistical data on the TOP 500 list [1] show that the variety of the HPC systems has been increasing. Scalar systems that consist of a large number of scalar processors perform highly parallel calculations for massively parallel applications [2]. Accelerator-type systems have been equipped with accelerators or co-processors such as GPUs (Graphics Processing Units) and Intel Xeon Phi processors. These accelerator-type systems can efficiently perform data parallel calculations by utilizing their high computational potential and memory bandwidth [3, 4].Vector systems that employ vector processors can calculate sets of data elements at the same time by utilizing powerful vector computation cores and their essentially high sustained memory bandwidths [5, 6].

In order to exploit the potential of various HPC systems, an application code should be appropriately written for effectively utilizing features of each HPC system. One of the most promising approaches is to employ directives that enable an application to effectively utilize the features. For example, OpenMP [7] provides a set of directives, called an OpenMP directive set, to enable a serial code to be executed by multithread parallel processing on a shared memory system. OpenACC [8] provides a set of directives, called an OpenACC directive set, to allow a code to be executed on an accelerator. Compiler-specific directive sets give a compiler additional information to use unique features of HPC systems and/or to encourage more effective analyses and aggressive compiler optimizations.

However, an effective way of using those directives strongly depends on both an application code and a target HPC system. To efficiently run one application on multiple HPC systems, various directives defined in different directive sets might be required to exploit the performances of those systems. Generally, an OpenMP directive set is used for a multiple core system, while an OpenACC directive set is used for an accelerator-type system. If the target HPC systems are both of the systems, different kinds of directive sets have to be maintained in one code; for example, both OpenMP and OpenACC directives are inserted into one application code. As a result, a large number of code lines would be spent for directives, which do not express the program behavior and are used only for performance. Thus, the maintainability and readability of the code decreases.

This paper proposes a directive translation approach that translates a special placeholder in an application code into different directives, depending on a target system. By utilizing a code translation framework, *Xevolver*, the special placeholder can be translated into any other directives for various HPC systems. Therefore, a single application code with special placeholders can be used for various HPC systems, which can keep the maintainability and readability of the original code.

The rest of this paper is organized as follows. Section 2 describes commonly used directive sets that exploit the potential of HPC systems. Section 3 proposes a directive translation approach that translates a special placeholder into any appropriate directives by using the Xevolver code translation framework. Section 4 demonstrates a translation of the special placeholder into an OpenMP directive for a target HPC system to clarify the effectiveness of the proposed approach. Section 5 gives concluding remarks and future work.

2 Utilization of Directive Sets for HPC Systems

In the field of HPC, directives are often used in various ways. For example, directives are adopted to allow parallel execution by multiple threads on a shared memory system, parallel execution on an accelerator, use of particular features of HPC systems, and so on. One of the reasons to use directives is that various functions offered by different directive sets are easily utilized by inserting directives into an existing application code. Furthermore, inserted directives can be ignored by compilers if the corresponding compiler function is disabled. All the directives are treated as comment lines. Moreover, an application code using directives can be incrementally developed. Even if a whole application code is not optimized and the optimization process is ongoing, the application can at worst execute. Therefore, an application developer can easily try using directives.

Another reason is that use of directives can avoid drastic code modification. Simply inserting directives into a code does not need to modify the structure of the original code. Thus, an application developer can easily continue to use the original code with directives.

However, effective use of directives for an application depends on the HPC system executing the application. As a result, various directives from multiple directive sets need to be used into one application code in order to maintain one unified version of the code. Figure 1 shows a code example, in which three kinds of directive sets, OpenMP, OpenACC, and compiler-specific directives, are used. In the code, OpenACC, OpenMP, and compiler-specific directives start with *!$acc*, *!$omp*, and *!cdir*,

```
71:!$omp parallel do &
72:!$omp& private(r_k,cp_j,cp_jmh,cp_jph,SQRTG_ijk,divvrq,diffrq)
73:!cdir novector
74:!$acc kernels                        &
..:!$acc ...                            &
80:!$acc present(rad, phi, phi_v, r2b1, rad_w)
81:!$acc loop gang
82:     do j = js,je
83:!cdir on_adb(sqrtg_w)
84:!$acc loop vector
85:       do i = is,ie
86:         sqrtg_w(i) = 1.0_DP/sqrtg(i,j)
87:       end do
88:!cdir outerunroll=4
99:!$acc loop gang
90:       do k = ks,ke
91:!cdir on_adb(sqrtg_w)
92:!$acc loop vector
93:         do i = is,ie
94:           r_k     = rad(k)
...
128:           end do
129:         end do
130:       end do
131:!$omp end parallel do
132:!$acc end kernels
```

Fig. 1 An application code with multiple directive sets

respectively. In this way, simultaneous use of various kinds of directive sets increases the number of code lines, and spoils the code maintainability and readability.

Therefore, keeping the maintainability and readability of an original code is strongly required even by applying directive-based optimization, especially in various HPC systems.

3   Directive Translation using *Xevolver* Code Translation Framework

This section proposes a directive translation approach that translates a special placeholder to different directives. This paper assumes that a special placeholder is used to specify the code line where one or more directives are potentially inserted for high performance.

A key idea of the proposed approach is to translate those placeholders to appropriate directives for individual HPC systems by using the *Xevolver* code translation framework. The Xevolver framework is a code translation framework that can define custom translation rules for any code modifications [9, 10]. The pre-defined and/or user-defined translation rules can be applied to a code. Because code modifications can be represented separately from an application code, the Xevolver framework can keep the maintainability of the original code. By defining translation rules for directive translation on the Xevolver framework, the special placeholders can be translated into directives for any target HPC systems.

Figure 2 shows an overview of the proposed directive translation approach. First, instead of OpenMP, OpenACC, and/or compiler-specific directives, a special place-

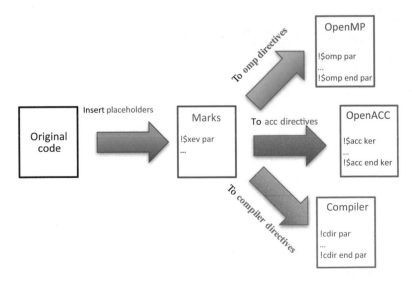

Fig. 2   An overview of directive translations

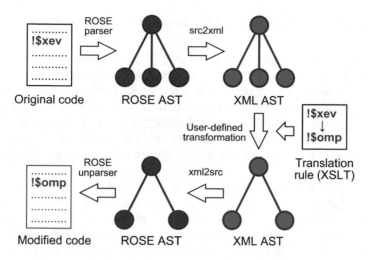

Fig. 3 An overview of directive translation using the Xevolver framework

holder is inserted into the original application code. Then, the Xevolver framework takes responsibility of translation of the special placeholder into appropriate directives. Considering the target HPC system, the translation is carried out based on user-defined translation rules. By defining multiple translation rules, each of which is corresponding to one HPC system, appropriate directive translation for individual systems can be realized. In the figure, the placeholder *!$xev* can be translated into OpenMP, OpenACC, or compiler-specific directives. As a result, one application code with a placeholder can be utilized for various HPC systems, which can keep the maintainability and readability of the original code.

Figure 3 shows an overview of directive translation by the Xevolver framework. First, an application code with a placeholder is parsed by using the ROSE compiler infrastructure [11], and then its AST (Abstract Syntax Tree) is converted to an XML document by the *src2xml* command of the Xevolver framework. An AST is represented as an XML document called an *XML AST*. The XML AST is translated into a new XML AST based on a translation rule. By considering code patterns and structures using AST information, more advanced directive translation can be realized than just text replacements using a *sed* command and so on. The translation rule can be described by XML data conversion format XSLT (XML Stylesheet Language Transformations) [12] or can also be generated from Fortran codes by using an Xevolver tool chain [13]. After the transformation of the original XML AST, the new XML AST is converted by the *xml2src* command of the Xevolver framework. Then, it is unparsed through the ROSE compiler infrastructure. Finally, an application code with translated directives is generated [10].

4   Translation of a Placeholder into OpenMP Directives

This section shows a case study of the translation of a special placeholder into
an OpenMP directive set. A kernel of an atmospheric simulation code [14, 15] is
used for the case study. Figure 4 shows the kernel of the simulation code, into
which a special placeholder is already inserted. The placeholder is inserted based
on compile information that is obtained by NEC SX Fortran compiler Rev.520 [16].
In Line 52, the placeholder *"!$xev sx_parallelizable"* is inserted because the SX
Fortran compiler could apply automatic parallelization for the loop body from Lines
53 to 83.

The inserted placeholder is translated into OpenMP directives by the proposed
approach. Figure 5 shows the translation rule to translate the placeholder into the
OpenMP do and *OpenMP end do* directives, which is written in XSLT. From Lines
1 to 3, the type of this document is defined. These lines indicate that this document is described in XSLT. Line 5 declares the variable *placeholder* that has *"!$xev
sx_parallelizable"*. Lines 7 to 9 specify that the following templates from Lines 11
to 37 are applied for whole AST. From Lines 12 to 17, all nodes in AST are copied
by the *copy* and *copy-of* elements. Since each XML element of *PreprocessingInfo*
represents a comment or a directive, Line 22 checks if each *PreprocessingInfo* element represents a placeholder. Thus, Line 22 looks for the *PreprocessingInfo* whose
text starts with *"!$xev sx_parallelizable"*. Finally, the directives *"!$omp parallel
do"* and *"!$omp end parallel do"*, target directives of the placeholder, are inserted
by Lines 23 to 25 and Lines 26 to 28, respectively. In order to insert *"!$omp end
parallel do"* into an appropriate position, the end of the loop body has to be figured
out. Then, right after the loop body, the *OpenMP end* directive needs to be inserted.
Only the value of the *pos* attribute in Line 26 is set to 3. As a result, the *OpenMP
end* directive is appropriately inserted after the loop body. Because it is difficult for
text replacements to find the appropriate position and to insert such directives, AST
translation is essential.

Figure 6 shows the kernel with the translated OpenMP directives. By using the
Xevolver framework and the translation rule, the OpenMP directives are successfully
inserted into Lines 53 and 85 in the code. Therefore, this case study demonstrates
that a special placeholder can be translated for a particular directive. As a result, an

Fig. 4 An atmospheric
simulation kernel with a
placeholder

```
71:!$xev sx_parallelizable
72:     do j = js,je
73:       do i = is,ie
74:         sqrtg_w(i) = 1.0_DP/sqrtg(i,j)
75:       end do
76:       do k = ks,ke
87:         do i = is,ie
88:           r_k     = rad(k)
...:
111:          end do
112:        end do
113:      end do
```

```
 1:<?xml version="1.0" encoding="UTF-8"?>
 2:<xsl:stylesheet version="1.0"
 3:  xmlns:xsl="http://www.w3.org/1999/XSL/Transform">
 4:
 5:  <xsl:variable name="placeholder" select="'!$xev_sx_parallelizable'"/>
 6:
 7:  <xsl:template match="/">
 8:    <xsl:apply-templates />
 9:  </xsl:template>
10:
11:  <!-- copy nodes -->
12:  <xsl:template match="*">
13:    <xsl:copy>
14:      <xsl:copy-of select="@*" />
15:      <xsl:apply-templates />
16:    </xsl:copy>
17:  </xsl:template>
18:
19:  <!-- remove placeholders -->
20:  <xsl:template match="PreprocessingInfo">
21:    <xsl:choose>
22:      <xsl:when test="starts-with(text(),$placeholder)">
23:        <PreprocessingInfo type="3" pos="2">
24:          !$omp parallel do
25:        </PreprocessingInfo>
26:        <PreprocessingInfo type="3" pos="3">
27:          !$omp end parallel do
28:        </PreprocessingInfo>
29:      </xsl:when>
30:      <xsl:otherwise>
31:        <xsl:copy>
32:          <xsl:copy-of select="@*" />
33:          <xsl:apply-templates />
34:        </xsl:copy>
35:      </xsl:otherwise>
36:    </xsl:choose>
37:  </xsl:template>
38:</xsl:stylesheet>
```

Fig. 5 The translation rule to translate the placeholder into OpenMP directives

Fig. 6 An atmospheric simulation kernel with translated OpenMP directives

```
47:!$omp parallel do
48:DO j = js, je
49:DO i = is, ie
50:sqrtg_w(i) = 1.0_DP / sqrtg(i,j)
51:END DO
52:DO k = ks, ke
53:DO i = is, ie
54:r_k = rad(k)
..:
77:END DO
78:END DO
79:END DO
80:!$omp end parallel do
```

application developer can concentrate on maintaining only the original code with a special placeholder.

By defining multiple translation rules in advance, each of which is corresponding to one HPC system, the placeholder can be translated into an appropriate directive for any target HPC systems.

5 Conclusions

This paper proposes a directive translation that can translate a special placeholder into directives for various HPC systems by using the Xevolver code translation framework. To exploit the potential of various HPC systems with only one application code, various directives of multiple directive sets tend to be inserted because an appropriate way of using directives for each HPC system is different. To avoid decreasing the maintainability and readability due to multiple kinds of directive sets, the proposed approach inserts a special placeholder instead of inserting actual directives. Then, the placeholder is translated into any directives for HPC systems by using the Xevolver code translation framework. By defining multiple translation rules in advance, each of which is corresponding to one HPC system, the placeholder can be translated into an appropriate directive for any target HPC systems. Finally, the demonstration illustrates that translation from a special placeholder into OpenMP directives can be successfully achieved. Especially, because the information about code structure is essential to insert *OpenMP end* directives, it is clarified that the translation through AST is required for the proposed directive translation approach. It also clarified that as a translation rule can be written in an external file, the proposed approach can keep the maintainability and readability of the original code.

Our future work includes translation of a special placeholder into other directives to clarify the effectiveness of the proposed directive translation approach. Furthermore, the performance with the translated directives needs to be clarified.

Acknowledgements This research was partially supported by Core Research of Evolutional Science and Technology of Japan Science and Technology Agency (JST CREST) "An Evolutionary Approach to Construction of a Software Development Environment for Massively-Parallel Heterogeneous Systems". This research uses SX-ACE in Cyberscience Center of Tohoku University.

References

1. Top500 supercomputer sites. http://www.top500.org/
2. Hasegawa, Y., Iwata, J.I., Tsuji, M., Takahashi, D., Oshiyama, A., Minami, K., Boku, T., Shoji, F., Uno, A., Kurokawa, M., Inoue, H., Miyoshi, I., Yokokawa, M.: First-principles calculations of electron states of a silicon nanowire with 100,000 atoms on the k computer. In: Proceedings of 2011 International Conference for High Performance Computing, Networking, Storage and Analysis, SC '11, pp. 1:1–1:11 (2011). doi:10.1145/2063384.2063386
3. Rahimian, A., Lashuk, I., Veerapaneni, S., Chandramowlishwaran, A., Malhotra, D., Moon, L., Sampath, R., Shringarpure, A., Vetter, J., Vuduc, R., Zorin, D., Biros, G.: Petascale direct numerical simulation of blood flow on 200k cores and heterogeneous architectures. In: Proceedings of the 2010 ACM/IEEE International Conference for High Performance Computing, Networking, Storage and Analysis, SC '10, pp. 1–11 (2010). doi:10.1109/SC.2010.42
4. Shimokawabe, T., Aoki, T., Takaki, T., Endo, T., Yamanaka, A., Maruyama, N., Nukada, A., Matsuoka, S.: Peta-scale phase-field simulation for dendritic solidification on the tsubame 2.0 supercomputer. In: Proceedings of 2011 International Conference for High Performance Computing, Networking, Storage and Analysis, SC '11, pp. 3:1–3:11 (2011). doi:10.1145/2063384.2063388

5. Dunigan Jr., T.H., Vetter, J.S., White III, J.B., Worley, P.H.: Performance evaluation of the cray x1 distributed shared-memory architecture. IEEE Micro **25**(1), 30–40 (2005). doi:10.1109/ MM.2005.20

6. Soga, T., Musa, A., Shimomura, Y., Egawa, R., Itakura, K., Takizawa, H., Okabe, K., Kobayashi, H.: Performance evaluation of nec sx-9 using real science and engineering applications. In: Proceedings of the Conference on High Performance Computing Networking, Storage and Analysis, SC '09, pp. 28:1–28:12 (2009). doi:10.1145/1654059.1654088

7. The OpenMP API specification for parallel programming. http://openmp.org/

8. OpenACC directives for accelerometers. http://www.openacc-standard.org/

9. Komatsu, K., Egawa, R., Hirasawa, S., Takizawa, H., Itakurayz, K., Kobayashi, H.: Migration of an atmospheric simulation code to an openacc platform using the xevolver framework. In: Proceedings of the Third International Symposium on Computing and Networking, pp. 528–534 (2015)

10. Takizawa, H., Hirasawa, S., Hayashi, Y., Egawa, R., Kobayashi, H.: Xevolver: An XML-based code translation framework for supporting HPC application migration. In: 21st International Conference on High Performance Computing (HiPC 2014), pp. 1–11 (2014). doi:10.1109/ HiPC.2014.7116902

11. Quinlan, D., Liao, C.: The ROSE source-to-source compiler infrastructure. In: Cetus Users and Compiler Infrastructure Workshop, in conjunction with PACT 2011 (2011)

12. XSL transformations (XSLT) version 2.0. https://www.w3.org/TR/xslt20/

13. Suda, R., Takizawa, H., Hirasawa, S.: Xevtgen: fortran code transformer generator for high performance scientific codes. In: Proceedings of the Third International Symposium on Computing and Networking, pp. 528–534 (2015)

14. Takahashi, K., Azami, A., Tochihara, Y., Kubo, Y., Itakura, K., Goto, K., Kataumi, K., Takahara, H., Isobe, Y., Okura, S., Fuchigami, H., Yamamoto, J.i., Takei, T., Tsuda, Y., Watanabe, K.: World-highest resolution global atmospheric model and its performance on the Earth Simulator. In: International Conference for High Performance Computing, Networking, Storage and Analysis (SC), 2011, pp. 1–12 (2011)

15. Takahashi, K., Onishi, R., Sugimura, T., Baba, Y., Goto, K., Fuchigami, H.: Seamless simulations in climate variability and HPC. In: Resch, M., Roller, S., Benkert, K., Galle, M., Bez, W., Kobayashi, H. (eds.) High Performance Computing on Vector Systems 2009, pp. 199–219. Springer, Berlin (2010)

16. Komatsu, K., Egawa, R., Takizawa, H., Kobayashi, H.: A compiler-assisted OpenMP migration method based on automatic parallelizing information. In: Proceedings of 29th International Supercomputing Conference, vol. 8488, pp. 450–459 (2014)

An Automatic Performance Tracking System for Large-Scale Numerical Applications

Shoichi Hirasawa, Hiroyuki Takizawa and Hiroaki Kobayashi

Abstract In this work, we propose an Automatic Performance Tracking System for analyzing the changes in execution performance and finding the source code modifications that cause the degradation of performance portability. The proposed system works in order to support evolving a large-scale numerical application while maintaining its performance portability across multiple target computing systems. By evaluating the performance of an application on every computing system, the proposed system helps application developers find the source code modifications that degrade the execution performance on a computing system. The proposed system also retrieves multiple versions of an application from its code repository, and automatically executes them on a newly added computing system. As a result, application developers are able to analyze how the source code modifications in the past affect the performance on the new computing system. Based on the evaluation results, the application developers can review the source code changes to improve the performance portability of the HPC application through the system.

1 Introduction

Multiple types of computing systems and tool chains are widely used these days. High-performance computing (HPC) applications sometimes need to migrate to new target computing systems because of their long software life cycles. The burden of migrating HPC applications to new target computing systems is usually heavy because of the large code sizes of such applications.

To alleviate the heavy cost of the migration, the code of an application should be maintained in such a way as to be able to execute in high performance on multiple

S. Hirasawa (✉) · H. Takizawa · H. Kobayashi
Graduate School of Information Sciences, Tohoku University, Sendai, Japan
e-mail: hirasawa@sc.cc.tohoku.ac.jp

H. Takizawa
e-mail: takizawa@tohoku.ac.jp

H. Kobayashi
e-mail: koba@tohoku.ac.jp

© Springer International Publishing AG 2016   119
M.M. Resch et al. (eds.), *Sustained Simulation Performance 2016*,
DOI 10.1007/978-3-319-46735-1_10

computing systems. In this work, the capability of an HPC application to achieve high performance on different types of computing systems is defined as *performance portability*. If an application code has high performance portability, it is expected to easily migrate the application to a new target computing system.

HPC applications are usually optimized only for a small number of computing systems to increase their execution performances. When optimizations are applied to an application with consideration only for specific computing systems, the execution performance on other types of computing systems may degrade. As a result, optimization efforts taking a long time for a small number of computing systems may lead to degrading performance portability of the application.

Generally, code optimizations for a specific computing system may degrade the performance of the application on another system. Thus, it is necessary to prevent applying such optimizations so as to keep the performance portability high. The degradation of performance portability can be detected by finding out performance degradation of the application on a computing system. To find out the performance degradation, execution performance on every computing system needs to be obtained, tracked, and compared. Although unit testing frameworks [1] and automatic bug detection methods [2, 3] have been proposed, to the best of our knowledge, there is no performance tracking system to maintain high performance portability of HPC applications.

In this work, an Automatic Performance Tracking System (APTS) that supports maintaining high performance portability of HPC applications is designed and developed. The APTS finds the changes of a source code that decrease the performance portability of the application. Because of the complexity of current computing systems, it is difficult to model and predict the execution performances of HPC applications on their target computing systems. With the APTS, execution performances of applications are obtained by actually executing them on every target computing system.

This paper is organized as follows. Section 2 discusses the performance portability of HPC applications. Section 3 proposes the APTS. Section 4 evaluates the APTS and Sect. 5 provides conclusions and future work.

2 Performance Portability of HPC Applications

In this work, computer systems used for developing and executing HPC applications are categorized into the following three types. Figure 1 shows the computer systems that are considered to be used in developing, building, and executing the applications.

1. Development systems: computer systems that are used to edit application source codes.
2. Building systems: computer systems that are used to compile source codes and build execution binaries of applications.
3. Execution systems: computer systems that are used to execute applications.

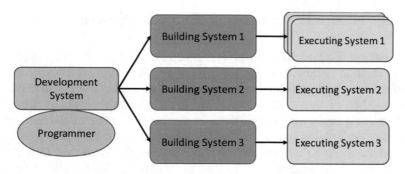

Fig. 1 Target computers for application development

When a programmer builds an application on the same system as the development system, the building environment such as compiling tool chains needs to be installed on the development system. Multiple building environments need to be installed on the development system when the application is developed considering multiple execution systems. However, installing all building environments on the development system is not always possible especially when software licences for production compilers are not available to programmers. In this work, therefore, an environment of multiple building systems and one development system is supposed to be used for developing applications. In the following subsections, programmer's burdens that can potentially be reduced by using tools are discussed.

2.1 Finding the Cause of Performance Degradation on Execution Systems

The source code of an HPC application tends to be large because many algorithms and optimizations are sometimes added during its long software life cycles. When a programmer finds a performance degradation of such an application on an execution system, the programmer needs to find the cause of degrading the performance of the application from the source code. The burden is usually heavy because of the large size of the source code of the application. To alleviate the burden, a system that notifies the cause of performance degradation on an execution system is useful.

Additionally, new execution systems are sometimes added to the target execution systems of an application because the life cycles of HPC applications tend to be longer than those of current execution systems. The programmer needs to know code modifications in the past to find the cause of performance degradation when a new execution system is added to the target systems.

2.2 Application Performance on Execution Systems

It is usually difficult to statically predict execution performance of an application on an execution system. It is because current execution systems have become too complex to create accurate performance models. As a result, an application needs to be executed on the target execution systems to evaluate the execution performance.

To evaluate execution performance, the source code needs to be built on the corresponding building systems of the target execution systems because execution performance of an HPC application depends on compilers and their optimization flags. With multiple development systems, programmers need to build and execute applications while editing one source code multiple times on the development systems.

2.3 Source Codes Synchronization Among Multiple Building Systems

When multiple building systems are used in addition to the development system as in Fig. 1, source codes need to be synchronized among them. Currently, programmers need to synchronize them manually with multiple tools while editing the source code of the application. This task is tedious and error-prone. Therefore, a tool that automatically synchronizes source codes among the development system and building systems can potentially ease the burden.

3 An Automatic Performance Tracking System

In this work, an Automatic Performance Tracking System (APTS) is proposed. The APTS is executed on the development system. In this work, it is assumed that a target application already has a build script and also an execution script with input data. Therefore, using the scripts, the APTS can build and execute the application on every execution system for performance evaluation and result verification.

3.1 Overview of the Automatic Performance Tracking System

The APTS tracks the changes of execution performance along with the modifications on an application code. Execution performances are automatically profiled by actually executing the application on its target execution systems. With the execution performances, the APTS finds the source code modifications that are causes of degrading the execution performance on an execution system. A programmer is able to use the found source code modifications to improve the performance portability

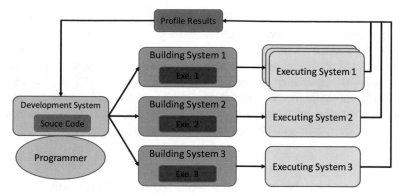

Fig. 2 Development framework of the APTS

of the application. For a programmer, it is easy to migrate an application to a new execution system if the application has high performance portability. The APTS has the following functions to help a programmer develop an application while keeping high performance portability (Fig. 2).

1. A function to track the changes of execution performance along with source code modifications.
2. A function to automatically build and execute applications on their target execution systems.
3. A function to automatically synchronize source codes among all building systems.

3.2 Performance Tracking Function Along with Source Code Modifications

While developing and optimizing an application, the modification to improve the execution performance on one execution system may degrade the execution performance on another execution system. When a new system is added to the execution systems of an application, the execution performance on the new system might be too low compared to its peak performance. In such a case, the low execution performance might be due to a certain source code modification for performance optimization in the past. Note that the new execution system was not available at the time, and the programmer could not check if the modification degrades the performance until the new system becomes available. Therefore, the programmer is required to check if every source code modification in the past degrades the performances on the available execution systems by tracking the past source code modifications whenever a new system becomes available.

The APTS uses version controlling systems such as CVS [4] or Git [5] to automatically find the performance changes with the past source code modifications on the newly added execution system. When a new system is added to the execution systems, the APTS automatically retrieves the past source codes of the application from the version controlling system. Then, the APTS automatically profiles the execution performance on the new execution system for every past version of the source code. By comparing the execution performances of two neighboring versions, performance degradation on the execution system can be detected. With the neighboring version numbers, the programmer can inspect an actual cause of the performance degradation on the new execution system.

4 Evaluation of the Automatic Performance Tracking System

4.1 Evaluation Setups of Finding the Cause of Degrading Performance portability

In this evaluation, the APTS is implemented as a plug-in program of the Eclipse integrated development environment (Eclipse IDE). It is implemented with the Plug-in Development Environment (PDE), which is the standard development framework of plug-in programs for the Eclipse. `Eclipse 4.2.1 Build id:20121004-1855` is used for developing and executing the APTS.

The source codes synchronization function is implemented with the Secure Copy (scp) command. `OpenSSH_6.1p1` is used in the APTS. To build the source codes of the target application, `Makefile` and `make` command are used. The executable file of the application is launched using the Secure Shell (ssh) command on the target execution systems. The `time` command is used to obtain the execution performances.

The APTS is evaluated to check if it is able to find the source code modifications that are the causes to degrade the performance portability of an application. A real HPC application of the entire growth process of binary alloy nanopowders in thermal plasma synthesis [6] is used in this evaluation. Three building systems are used from one development system. All building systems are installed in Cyberscience Center of Tohoku University. The development system, which is a desktop PC (Intel Core i7-3930K 3.2 GHz, 16 GB Memory, SSD), is installed in another building of mechanical engineering in the same campus of Tohoku University. The specifications of building systems are shown in Table 1. Server 1, 2 and 3 are also used as execution systems corresponding to the building systems.

4.2 Results of Analysing the Degradation of the Performance Portability

The evaluation results are shown in Fig. 3. The horizontal axis indicates version numbers of the target application. The vertical axis on the left-hand side shows the speedup ratio from the execution time of the application code Version 1 running on Server 1. The vertical axis on the right-hand side shows the number of modified source code lines between a neighboring two versions.

The application has been optimized for Server 1 along with the version numbers. Hence, the performance of Server 1 increases with the version number. On the other hand, the performance degrades by changing from Version 5 to 6 on Server 2. The Tesla C2070 GPU of Server 2 is newer than the Tesla C1060 GPU of Server 1. From these results, it is observed that the change from Version 5 to 6 degrades the performance portability of the application.

As the execution performance does not degrade on Server 3, which has a newer K20 GPU than C2070, the modification between Versions 5 and 6 only degrades the execution performance on Server 2 among the three. The number of different source code lines between Versions 5 and 6 is 37.

In the evaluation results, it is shown that the APTS is able to limit the number of source code lines that cause the performance degradation on execution systems. In this particular evaluation, the APTS can successfully reduce the number of source

Table 1 Specifications of building systems and execution systems

System name	Linux ver.	CPU	GPU	CUDA
Server 1	2.6.18	Core i7 920 2.67GHz	Tesla C1060	5.0
Server 2	2.6.32	Core i7 930 2.8GHz	Tesla C2070	5.0
Server 3	2.6.18	Core i7 920 2.67GHz	Tesla K20c	5.0

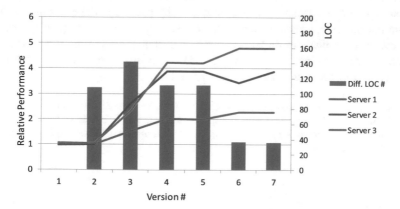

Fig. 3 Execution performances of application versions and line numbers of corresponding code difference

code lines that cause the performance degradation to 37. As a result, the APTS is able to support programmers to develop a large-scale application with high performance portability.

5   Conclusions and Future Work

In this paper, we have designed and implemented an Automatic Performance Tracking System (APTS). It automatically finds the version of an application, from which the performance is degraded on an execution system. It is implemented as a plug-in program of the Eclipse IDE. It has a function of transferring necessary files of an application to build machines. It then builds and executes the application to collect its execution performance on each execution system.

The APTS supports the development work of programmers by alleviating the burden of building and executing the application on multiple systems. It finds the version of an application code that degrades the execution performance on an execution system. As a result, the APTS helps a programmer maintain high performance portability of an application by keeping the execution performance high on multiple execution systems.

With the evaluation results, it has been shown that the APTS can successfully find the version of the real application that degrades the execution performance on an execution system. It has also been shown that the APTS can obtain the execution performances of the application on multiple execution systems by transferring and building the application on multiple building systems. With these functions, the manual work of performance evaluation necessary for programmers is automated and, as a result, the APTS is able to support the development work on maintaining high performance portability of HPC applications.

Realizing functions such as automatically evaluating the performance with profilers such as gprof and nvprof, obtaining the performance profile results, and reasoning the codes that degrade performance portability by providing the profiling results to the programmers are parts of our future work. Supporting batch queuing systems for executing applications on HPC computing systems is also important. Migrating the implementation for the code base of PTP [7] is also considered to provide the information of execution performance in the editor.

Acknowledgements The authors would like to thank Prof. Shigeta of Osaka University for allowing us to use the application. This work is partially supported by JST CREST "An Evolutionary Approach to Construction of a Software Development Environment for Massively-Parallel Heterogeneous Systems."

References

1. Zhu, H., Hall, P.A.V., May, J.H.R.: Software unit test coverage and adequacy. ACM Comput. Surv. **29**(4), 366–427 (1997)
2. Kim, S., Zimmermann, T., Pan, K., Whitehead, E.J.: Automatic identification of bug-introducing changes. In: 21st IEEE/ACM International Conference on Automated Software Engineering, 2006. ASE '06, pp. 81–90 (2006)
3. Williams, C.C., Hollingsworth, J.K.: Automatic mining of source code repositories to improve bug finding techniques. IEEE Trans. Soft. Eng. **31**(6), 466–480 (2005)
4. http://cvs.nongnu.org/. Cvs - concurrent versions system
5. http://gitscm.com/. Git - the fast version control system
6. Shigetam, M., Watanabe, T.: Growth model of binary alloy nanopowders for thermal plasma synthesis. J. Appl. Phys. **108**(4), 043306–043306–15 (2010)
7. Watson, G.R., Rasmussen, C.E., Tibbitts, B.R.: An integrated approach to improving the parallel application development process. In: IEEE International Symposium on Parallel Distributed Processing, 2009. IPDPS 2009, pp. 1–8, May 2009

Part II
Numerical Computations and Approach Towards Multi-physics Applications

A Case Study of Urgent Computing on SX-ACE: Design and Development of a Real-Time Tsunami Inundation Analysis System for Disaster Prevention and Mitigation

Hiroaki Kobayashi

Abstract In 2011, a huge earthquake named Great East-Japan Earthquake gave a serious damage in Japan, especially due to Tsunami inundation caused by the earthquake. After this terrible natural disaster, there is a growing concern about future big earthquakes and Tsunami disasters in Japan, and a demand for their prevention and mitigation is increasing. To react this high demand, we are designing and developing a realtime Tsunami inundation analysis system on a brand-new vector supercomputer SX-ACE installed at Tohoku University as a case study of urgent computing for earthquake and Tsunami disasters. In this article, we will present an overview of the system and its performance in the Nankai trough earthquake case.

1   Introduction

On March 11, 2011, the East-Japan earthquake occurred with a magnitude of 9.0 in the Pacific coast of Tohoku, which is a northern part of the mainland of Japan. A huge Tsunami with a height of more than 40 m triggered by the earthquake arrived at coastal cities of the Tohoku area 30 min after the earthquake, went into the inland up to 10 Km, and destroyed the cities. More than 18,000 victims (dead or missing) are due to the huge Tsunami.

After the great East-Japan earthquake, there is a growing concern about future big earthquakes and Tsunami disasters in Japan, and a demand for their prevention and mitigation is increasing, because there is a high probability of several big earthquakes expected in the very near future in the sea around Japan. For example, according to the report published by Cabinet Office, Government of Japan [1], the probability of the Nankai Trough earthquake in the next 30 years is 70 %, and the total number of deaths due to this huge earthquake would have been estimated to reach 320,000 in the worst case, with an economical loss of 220 trillion yen.

H. Kobayashi (✉)
Tohoku University, Sendai 980-8578, Japan
e-mail: koba@tohoku.ac.jp

© Springer International Publishing AG 2016

M.M. Resch et al. (eds.), *Sustained Simulation Performance 2016*,
DOI 10.1007/978-3-319-46735-1_11

To react such a high demand, we start considering our contribution to the society for prevention and mitigation of natural disasters by using our HPC resources and research achievements. Here we focuses on the following three points:

- Prompt responses to disasters to reduce damages such as warning evacuation from dangerous zones and rescuing survivors as soon as possible,
- Detailed and highly accurate analysis and forecasting of Tsunami inundation soon after a big earthquake that may trigger a huge Tsunami, and
- Enhancement of the social resiliency against natural disasters by precise simulation using HPC.

Under these considerations, we are designing and developing a real-time Tsunami inundation analysis system. This article presents an overview of the system and discusses the Tsunami simulation on the SX-ACE supercomputer, which is the brand-new vector supercomputer developed by NEC and has been installed in 2015 at Tohoku University [2].

2   A Real-Time Tsunami Inundation Analysis System

Figure 1 shows a configuration of the Tsunami inundation analysis system. The system has been designed to provide the information about the inundation in the coastal cities with a high resolution of 10-m grids within 20 min at the latest after an earthquake occurrence. To realize such an aggressive target of a highly accurate, 10-m mesh-level Tsunami inundation simulation within a short time period, we can successfully construct a coupled system of a real-time GPS-based earthquake observation system and an on-line Tsunami simulation on our SX-ACE system. The GPS-observation system monitors the land motion at more than 1,300 points in Japan in the 24/7 operation. By using real-time measured land-motion data after an earthquake, the fault model of the earthquake is estimated, and the necessary parameters for the Tsunami simulation are automatically generated. These parameters are transferred

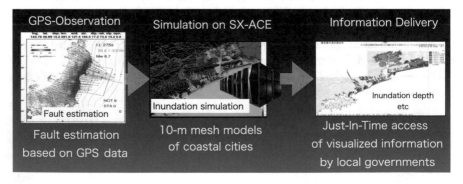

Fig. 1 Overview of the real-time Tsunami Inunation analysis system

to the Tsunami simulator on SX-ACE immediately via a network connecting them, and the simulation will be triggered soon after the parameters arrival. A job management system has specially be modified for the real-time Tsunami inundation analysis system. The job management system named NQS II has been enhanced to support urgent job prioritization so that the urgent job management function can execute the Tsunami inundation simulation job on the SX-ACE system at the highest priority, while immediately suspending other active jobs on the system. The suspended jobs automatically resume as soon as the Tsunami inundation simulation completes.

After the Tsunami simulation completes on the SX-ACE system, the results will be visualized on a city map, and delivered to officers of designated local governments. The visualized information provided includes Tsunami arrival time, maximum inundation depth, Tsunami level changes, estimation of damaged population, houses, and buildings, inundation start time, and maximum water level. All the visualized data can be available through the web-interface.

To realize a highly-accurate Tsunami inundation simulation within a reasonable time, we divide a computation domain in a hierarchical fashion. Figure 2 shows an

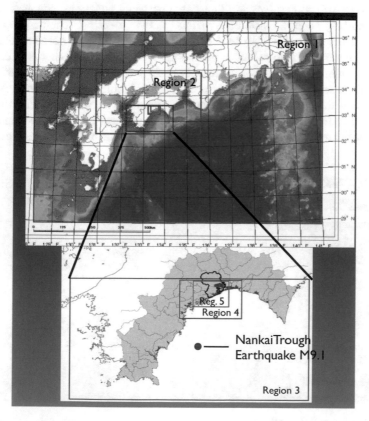

Fig. 2 Computation domain with a hierarchical decomposition for Kochi-City

Table 1 Grid size of computation domain for Kochi-City

Region	Grid size (m)	Num. of grid (X)	Num. of grid (Y)
1	810	1536	1020
2	270	1680	990
3	90	2292	1260
4	30	1782	1188
5	10	3504	2364

example of a hierarchical decomposition of the computation domain for Kochi city and its coastal area. Table 1 shows an actual grid sizes starting from a coarse grids of a 810-m mesh down to the finest grids of a 10-m mesh. With these multi-level grids, we implement the TUNAMI, (Tohoku University's Numerical Analysis Model for Investigating tsunami) code [3] for SX-ACE. The TUNAMI code has originally been developed by Tohoku University, and authorized by UNESCO and Japanese Government as the official code for Tsunami inundation analysis.

The TUNAMI code solves non-linear shallow water equations, and uses the staggered leap-frog 2D finite difference method as the numerical scheme [3]. Figure 3 shows a structure of the TUNAMI code. There are two high cost kernels in the TUNAMI code: one is to calculate mass conservation, and the other is to calculate motion equation. Therefore, we intensively apply several optimization techniques to these two kernels to exploit the potential of the SX-ACE supercomputer. As these kernels have the same structure of doubly nested loops, we vectorized the inner loop for efficient vector operations by the vector processor of the node, and parallelized the outer loop with MPI processes for efficient multi-node parallel processing.

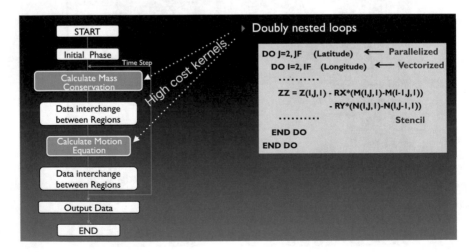

Fig. 3 Structure of TUNAMI code

In addition to the vectorization and parallelization, we developed several tuning techniques such as inlining subroutines, the optimization of I/O routines, and ADB tuning of stencil operations. Among these techniques, the ADB tuning is very important and effective to obtain a high sustained performance, because an on-chip memory named ADB (Assignable Data Buffer) is introduced to the vector processor of the SX-ACE to effectively provide data with a high locality to the vector pipes on a chip, without hight-cost off-chip memory accesses [4].

3 Performance Evaluation

In this section, we examine the performance of our implementation of the TUNAMI code on the SX-ACE system in comparison with a Xeon-based scalar-parallel system and the K computer. Table 2 summarizes the specifications of these systems. SX-9 and SX-ACE are vector systems and their advantages against scalar systems, LX 406 and the K-computer, are their higher memory bandwidths, when comparing the systems with the same peak performance, resulting in a higher system B/F, a ratio of a memory bandwidth to a peak performance. As many applications are memory-intensive, higher B/F is a key factor to improve their sustained performances. The Nankai trough case with a magnitude of 9.0 is used for the experiments.

The TUNAMI code was originally developed for Xeon-based scalar systems, however, our optimization techniques for its migration to the SX-ACE system lead to a significant performance improvement by a factor of 5.5 in comparison with the performance of the LX 406 system in the case of a single core execution as shown in Fig. 4. As a ratio of peak performances of two systems' cores is three, the computing efficiency, which is a ratio of the sustained performance to the peak performance, of the SX-ACE processor core is twice higher than that of the Xeon Ivy Bridge's core of the LX406.

The high memory bandwidth plays an important role to obtain higher efficiency rather than the peak performance in the execution of the TUNAMI code, because the TUNAMI code is a memory-intensive application. When comparing the performance of SX-ACE with that of SX-9, their execution times are almost same, even though the single core performance is only 62.5 % of SX-9's single core-performance. The higher computing efficiency of SX-ACE compared with SX-9 is the result of the enhancement of the memory subsystem of SX-ACE compared with that of SX-9, such as its shorter memory latency, an enlarged ADB with MSHR (Miss Status Handling Registers), a short-cut mechanism in chaining vector pipes, and out-of-order vector load operations [5].

Figure 5 shows the execution time of the simulation in the multi-node environment on SX-ACE, LX406 and the K computer. In the figure, numbers of marks mean the execution times of each system when changing the number of threads(cores). Figure 5 indicates that the performance of the SX-ACE system with 512 cores is equivalent

Table 2 Specification of evaluated systems

System	No. of sock- ets/node	Perf./socket (Gflop/s)	No. of cores/socket	Perf./core (Gflop/s)	Mem. BW/socket GB/s	On-chip mem	NW BW (GB/s)	Sys. B/F
SX-ACE	1	256	4	64	256	1MB ADB/core	2 × 4 IXS	1.0
SX-9	16	102.4	1	102.4	256	256KB ADB/core	2 × 128 IXS	2.5
LX 406 (Ivy Bridge)	2	230.4	12	19.2	59.7	256KB L2/core 30MB Shared L3	5 IB	0.26
K (SPARK64VIIIfx)	1	128	8	16	64	6MB Shared L2	5-50 Tofu NW	0.5

Fig. 4 Single-core performance

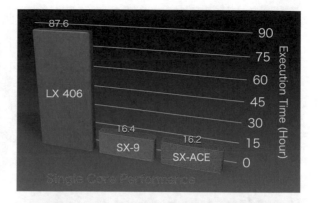

Fig. 5 Core scalability

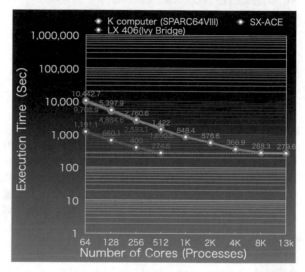

to that of 13 K cores of the K computer. This is because SX-ACE can achieve a high computation efficiency mainly thanks to its high memory bandwidth in cooperation with an on-chip memory named ADB. The high sustained performance of a single core also contributes to the decreasing number of MPI processes to obtain the certain level of the performance for the real-time Tsunami inundation simulation.

Figure 6 shows the timing chart of the entire process from the earthquake detection to the visualization of the simulation results. The process of the Tsunami inundation analysis system consists of two phases: coseismic fault estimation phase and tsunami inundation simulation phase. Using this system for the Nankai trough earthquake case in Japan with 6-h simulation of Tsunami's behavior, these two phases can complete in less than 8 min and 5 min, respectively. In addition, the simulation results can be visualized and sent to local governments within 4 min after the simulation. Thus, the system completes the Tsunami inundation analysis with a 10-m grids resolution within 20 min.

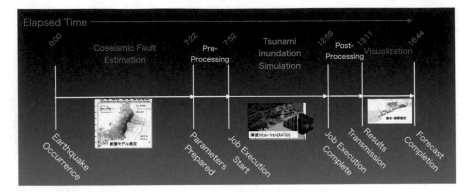

Fig. 6 Timing chart of the entire process

4 Summary

This article briefly described design and implementation of a real-time Tsunami inundation analysis system on the vector supercomputer SX-ACE. This is the world-first achievement of real-time Tsunami inundation analysis of a 6-h inundation behavior at the level of a 10-m mesh We can implement the system very efficiently by exploiting the potential of SX-ACE, and as a results, even with 512 nodes, we can complete a real-time simulation in 8 min after the occurrence of a big earthquake, which is equivalent to the performance of the K-computer with 13 K nodes. Therefore, we can successfully demonstrate that our SX-ACE system with the capability of the real-time Tsunami inundation simulation has a potential as a social infrastructure for the homeland safety like a weather forecasting system, in addition to as a research infrastructure for computation science and engineering applications.

Acknowledgements Many colleagues of Tohoku University, NEC and its related companies made a great contribution to this project, and in particular great thanks go to Professors Shunichi Koshimura, Ryota Hino, Yusaku Ohta, Visiting Professors Akihiro Musa and Yoichi Murashima, and Visiting Researchers Hiroshi Matsuoka and Osamu Watanabe, all from Tohoku University

References

1. Cabinet Office, Government of Japan. http://www.bousai.go.jp/jishin/nankai/nankaitrough_info.html
2. Kobayashi, H.: A new SX-ACE-based supercomputer system of Tohoku University. In: Sustained Simulation Performance 2015, pp. 3–151. Springer (2015)
3. Koshimura, S., et al.: Developing fragility functions for tsunami damage estimation using numerical model and post-tsunami data from Banda Aceh, Indonesia. Coastal Eng. J. 243–273 (2009)
4. Momose, S.: SX-ACE, Brand-new vector supercomputer for higher sustained performance I. In: Sustained Simulation Performance 2014, pp. 199–214. Springer (2014)
5. Momose, S.: Next generation vector supercomputer for providing higher sustained performance. In: COOL Chips 2013 (2013)

CFD/CAA Simulations on HPC Systems

Michael Schlottke-Lakemper, Fabian Klemp, Hsun-Jen Cheng,
Andreas Lintermann, Matthias Meinke and Wolfgang Schröder

Abstract In this paper, a highly scalable numerical method is presented that allows
to compute the aerodynamic sound from a turbulent flow field on HPC systems.
A hybrid CFD-CAA method is used to compute the flow and the acoustic field, in
which the two solvers are running in parallel to avoid expensive I/O operations for
the acoustic source terms. Herein, the acoustic perturbation equations are solved
by a high-order discontinuous Galerkin scheme using the acoustic source terms
obtained from an approximate solution of the Navier-Stokes equations. Both solvers
run simultaneously and operate on differently refined hierarchical Cartesian grids.
This direct-hybrid method is validated by monopole and pressure pulse simulations
and is used for performance measurements on current HPC systems. The results
highlight the limitations of classic hybrid methods and show that the new approach
is suitable for highly parallel simulations.

M. Schlottke-Lakemper (✉) · A. Lintermann
Jülich Aachen Research Alliance - High Performance Computing, RWTH Aachen University,
Aachen, Germany
e-mail: m.schlottke-lakemper@aia.rwth-aachen.de

A. Lintermann
e-mail: lintermann@jara.rwth-aachen.de

F. Klemp · H.-J. Cheng · M. Meinke · W. Schröder
Institute of Aerodynamics, RWTH Aachen University, Aachen, Germany
e-mail: f.klemp@aia.rwth-aachen.de

H.-J. Cheng
e-mail: h-j.cheng@aia.rwth-aachen.de

M. Meinke
e-mail: m.meinke@aia.rwth-aachen.de

W. Schröder
e-mail: office@aia.rwth-aachen.de

© Springer International Publishing AG 2016
M.M. Resch et al. (eds.), *Sustained Simulation Performance 2016*,
DOI 10.1007/978-3-319-46735-1_12

1 Introduction

One of the major challenges of today's aircraft development is noise reduction, which is also one of the central aims in European aircraft policy. The perceived noise levels of flying aircraft are to be reduced until 2050 by 65 % compared to the year 2000 [25]. Many sound-generating components of aircraft need to be assessed in sufficient detail to be able to improve their design, such as the optimization of the jet nozzle geometry to lower noise emissions at take-off without sacrificing the thrust efficiency. To achieve such optimizations, efficient, fully parallelized algorithms are needed to predict the flow field and the far-field noise of jet engines.

A hybrid method combining large-eddy simulation (LES) with computational aeroacoustics (CAA) for large-scale aeroacoustics simulations has been successfully applied in [7, 18]. It uses LES to determine the turbulent flow field for external flow configurations. From this solution, noise-generating source terms are extracted and used in a CAA simulation, where the acoustic field is predicted using the acoustic perturbation equations (APE) [6]. This scheme has been applied successfully to different problems in computational aeroacoustics, such as trailing edge noise [7], jet noise [12], or combustion noise [4, 11]. However, it suffers from the exchange of large data volumes for the acoustic source terms via I/O operations, which limits the efficiency of such a two-step approach especially on high-performance computing (HPC) systems.

To circumvent this bottleneck, the direct-hybrid method presented in this work combines the LES and CAA solvers in a single framework such that both solvers can run in parallel. The LES solver used for the prediction of the flow field is based on a finite-volume method, while the CAA approach makes use of a high-order discontinuous Galerkin (DG) method to solve the APE for the acoustic field. DG methods were first described by Reed and Hill [24] and were subsequently applied to various physical problems, such as incompressible and compressible flow [2, 22], magnetohydrodynamics [29], and aeroacoustics [1, 3].

The LES and CAA computations are performed on a joint Cartesian mesh. Based on a coloring scheme, cells are associated with different weights for the LES and CAA solution and a space-filling curve is used for the domain decomposition. The coupling mechanism between both simulations only requires memory transfer operations. That is, no additional communication between the subdomains is necessary, leading to an efficient algorithm to be used on massively parallel systems. Furthermore, this direct-hybrid approach allows a more fine-grained control over the coupling process itself, since the LES results are not obtained separately from the acoustic field anymore. This means that, e.g., the time step size or the grid size can be adapted during the simulation to account for time-dependent changes in the resolution requirements of both solvers, enabling in situ optimizations of the simulation process.

In this paper, the coupling approach for the direct-hybrid LES-CAA simulation is presented and results for performance measurements are shown. A CAA code is developed and integrated with an existing LES solver. After the governing equations are introduced in Sect. 2, the numerical methods are described in Sect. 3. In Sect. 4, the coupling strategy is discussed in detail. The CAA solver is validated in Sect. 5, before it is used for strong scaling experiments on two state-of-the-art HPC systems. In Sect. 6, the presented methods and the obtained results are summarized.

2   Governing Equations

In this hybrid CFD-CAA method, two sets of governing equations are utilized. One solely describes the generation and propagation of acoustic waves, while the other set of equations predicts the physics of the underlying flow field. Here, the acoustic perturbation equations are used for the acoustic field and the Navier-Stokes equations for the flow field. Both are briefly summarized in the following.

2.1   Navier-Stokes Equations

The Navier-Stokes equations in non-dimensional, conservative form are given by

$$\frac{\partial \rho}{\partial t} + \nabla \left(\rho \boldsymbol{u} \right) = 0,$$

$$\frac{\partial \rho \boldsymbol{u}}{\partial t} + \nabla \left(\rho \boldsymbol{u}\boldsymbol{u} + p + \frac{\tau}{\text{Re}_0} \right) = 0, \tag{1}$$

$$\frac{\partial \rho e}{\partial t} + \nabla \left((\rho e + p)\boldsymbol{u} + \frac{1}{\text{Re}_0}(\tau \boldsymbol{u} + \boldsymbol{q}) \right) = 0.$$

The quantity ρ represents the fluid density, $\boldsymbol{u}$ the velocity vector, and e the total specific energy. The system in Eq. (1) is closed by the definition of the total specific energy for a perfect gas

$$\rho e = \frac{p}{\gamma - 1} + \frac{1}{2}\rho(\boldsymbol{u} \cdot \boldsymbol{u}), \tag{2}$$

where p is the pressure and γ is the specific heat ratio. For non-dimensionalization, the stagnation state is employed, which is denoted by the subscript 0. The Reynolds number based on the stagnation state is defined by

$$\text{Re}_0 = \frac{\rho_0 c_0 L}{\mu_0}, \tag{3}$$

where L is a reference length and ρ_0, c_0, and μ_0 are the stagnation density, the speed of sound, and the dynamic viscosity. A Newtonian fluid is assumed such that the components τ_{ij} of the stress tensor τ can be written as

$$\tau_{ij} = -2\mu S_{ij} + \frac{2}{3}\mu S_{ij}\delta_{ij}, \tag{4}$$

where $S_{ij} = \frac{1}{2}\left(\frac{\partial u_i}{\partial x_j} + \frac{\partial u_j}{\partial x_i}\right)$ is the rate of strain tensor. The dynamic viscosity μ is calculated by using Sutherland's law and the vector of heat conduction q is determined by Fourier's law

$$q = -\frac{k}{\Pr(\gamma - 1)}\nabla T, \tag{5}$$

where T is the static temperature. The Prandtl number is defined with the specific heat at constant pressure c_p by $\Pr = \frac{\mu_0 c_p}{k_0}$. For a constant Prandtl number, the relation $k(T) = \mu(T)$ holds for the thermal conductivity.

2.2 Acoustic Perturbation Equations

The acoustic perturbation equations (APE) were introduced in [6] and are used to predict the acoustic field for flow-induced noise. They are derived from the linearized Euler equations and modified to retain only acoustic modes without generating vorticity or entropy modes. Neglecting all viscous, non-linear and entropy-related contributions, the APE-4 system reads [6]

$$\frac{\partial u'}{\partial t} + \nabla\left(\bar{u} \cdot u'\right) + \nabla\left(\frac{p'}{\bar{\rho}}\right) = q_m, \tag{6}$$

$$\frac{\partial p'}{\partial t} + \bar{c}^2 \nabla \cdot \left(\bar{\rho} u' + \bar{u}\frac{p'}{\bar{c}^2}\right) = 0, \tag{7}$$

where the source term q_m is the linear Lamb vector

$$q_m = -(\omega \times u)' = -(\omega' \times \bar{u} + \bar{\omega} \times u'), \tag{8}$$

with ω as the vorticity vector. The variables of the APE are perturbed quantities denoted by prime $(\cdot)'$ and are defined by $\phi' := \phi - \bar{\phi}$, where the bar $(\bar{\cdot})$ denotes time-averaged quantities.

In the present work, the non-dimensional form of Eqs. (6) and (7) is used. As for the Navier-Stokes equations, the stagnation state is used for the definition of reference values. Furthermore, it is assumed here that the time-averaged values for the speed of sound and density are constant and equal to the stagnation state, i.e., $\bar{c} = c_0$ and $\bar{\rho} = \rho_0$, which is only valid in the low-Mach number regime. By using the following non-dimensional variables,

$$\tilde{t} = \frac{t c_0}{L}, \qquad \tilde{x} = \frac{x}{L}, \qquad \tilde{u} = \frac{u}{c_0}, \qquad \tilde{p} = \frac{p}{\rho_0 c_0^2}, \tag{9}$$

the APE can be written as

$$\frac{\partial \tilde{\boldsymbol{u}}'}{\partial \tilde{t}} + \tilde{\nabla}(\tilde{\bar{\boldsymbol{u}}} \cdot \tilde{\boldsymbol{u}}' + \tilde{p}') = \tilde{\boldsymbol{q}}_m, \tag{10}$$

$$\frac{\partial \tilde{p}'}{\partial \tilde{t}} + \tilde{\nabla} \cdot (\tilde{\boldsymbol{u}}' + \tilde{\bar{\boldsymbol{u}}}\, \tilde{p}') = 0. \tag{11}$$

The non-dimensional source term is given by $\tilde{q}_m = \frac{q_m}{\tilde{c}_0^2/L}$. For convenience, in the following discussion the tilde is dropped from the non-dimensional quantities.

3   Numerical Methods

In this section, the meshing process and the domain decomposition are outlined. Furthermore, the numerical methods for the acoustic perturbation equations and the Navier-Stokes equations are briefly described.

3.1   Hierarchical Mesh Topology

Both the LES solver and the CAA solver operate on a joint hierarchical Cartesian mesh. The cells of the grid are organized in a tree structure (2D: quadtree, 3D: octree), with parent-child relationships between different levels and neighbor relationships within a level. The discretization process follows the method described in [21] and starts with a single square/cube cell which encloses the whole computational domain. This zero-level cell is then refined uniformly until the desired refinement level is reached (see Fig. 1a). A cell to be refined is isotropically subdivided into 2^d square/cube cells, with d being the number of spatial dimensions and with the original cell becoming the parent cell of the new child cells. Individual regions of

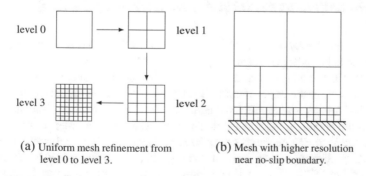

(a) Uniform mesh refinement from level 0 to level 3.

(b) Mesh with higher resolution near no-slip boundary.

Fig. 1 Cell refinement for a hierarchical Cartesian grid in 2D

Fig. 2 Domain partitioning
on two domains with four
subtrees starting at level l_α

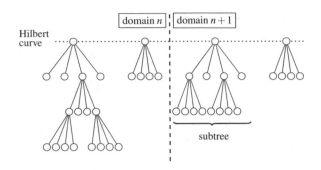

the mesh can be further refined to meet resolution requirements, e.g., in areas with
small-scale physical features such as wall-bounded shear layers or to accurately
resolve boundaries (see Fig. 1b). A smoothing algorithm ensures that the level difference between neighboring cells does not exceed one, i.e., each cell has at most
2^{d-1} neighbor cells in each spatial direction. Special treatment is necessary for cells
that are intersected by the body geometry. In this paper, only non-intersected cells
are considered. During grid generation, the zero-level cell is homogeneously refined
to a minimum level l_α and all coarser cells are discarded [21]. These cells at level
l_α become the roots of their subtrees and are further subdivided until the required
refinement level is reached.

For the domain decomposition, a Hilbert space-filling curve [26] is used to map
the grid at level l_α to the interval [0, 1]. Each cell at level l_α is assigned a load that
depends on the number of cells in its subtree and on the type of the cells, i.e., whether
they are LES or CAA cells. Load balancing is achieved by taking into account these
load values when distributing the cells among the processes and for each l_α cell the
entire subtree is placed on the same rank (see Fig. 2). By consecutively placing l_α
cells and their subtrees on the MPI ranks according to their position on the Hilbert
curve, spatial compactness is ensured, reducing the overall communication cost.

3.2 Discontinuous Galerkin Approximation of the APE

A discontinuous Galerkin spectral element method (DGSEM) is used to determine
the acoustic field. In Kopriva et al. [19], the DGSEM was proposed and has been used
extensively [9, 17]. Since it was derived for quadrilateral/hexahedral mesh elements,
it is well-suited for the use on hierarchical Cartesian grids. Furthermore, its compact
formulation allows a very efficient parallelization, when explicit time stepping is
used, and the parallel efficiency is independent of the chosen order of the scheme.

Since the DGSEM elements correspond to cells in a finite-volume context, the
words cell or element will be used interchangeably. In the following, the main components of the DGSEM are outlined. First, the system of equations is mapped to
a reference element for efficiency reasons. The derivation of the DG formulation

then starts with the weak formulation, choosing Lagrange polynomials to represent the solution within each element. This gives rise to an integral equation, which is approximately solved using Gauss quadrature. Finally, the discrete DG operator is integrated in time using a Runge-Kutta scheme.

A general system of hyperbolic conservation equations in three dimensions reads

$$\frac{\partial U}{\partial t} + \nabla \cdot f(U) = 0, \tag{12}$$

where $U = U(x, t)$ is the vector of conservative variables $\{u_i\}_{i=1}^{n_v}$ and f is the flux vector. For efficiency reasons, the differential equation is mapped to a reference element E, which is in three dimensions given by a cube of size $[-1, 1] \times [-1, 1] \times [-1, 1]$. Introducing the reference coordinate vector $\boldsymbol{\xi} = (\xi^1, \xi^2, \xi^3)^\mathsf{T}$, the final transformed equation reads [17]

$$\hat{J}U_t + \nabla_\xi \cdot f = 0, \tag{13}$$

where $\hat{J}$ is the Jacobian, which for cube-to-cube transformations is just $\frac{h}{2}$, h being the side length of the cube, and U_t is the time derivative of the vector of conservative variables.

The derivation of the DG method starts with the weak form of the equation. Therefore, Eq. (13) is multiplied by a test function $\phi = \phi(\boldsymbol{\xi})$ and integrated over the reference element E

$$\int_E \left(\hat{J}U_t + \nabla_\xi \cdot f\right) \phi \, d\boldsymbol{\xi} = 0. \tag{14}$$

Using integration by parts on the flux term, the weak formulation of the differential equation is obtained

$$\int_E \hat{J}U_t \phi \, d\boldsymbol{\xi} + \int_{\partial E} (f \cdot n)^* \phi \, ds - \int_E f \cdot \nabla_\xi \phi \, d\boldsymbol{\xi} = 0, \tag{15}$$

where n is the surface normal vector in the reference system. Similar to the finite-volume approach, the value for the normal flux $f \cdot n$ is not uniquely defined on the element boundaries ∂E, since the solutions in the left U^- and right U^+ elements are discontinuous. Therefore, a numerical flux $(f \cdot n)^* = g(U^+, U^-, n)$ is chosen that combines values from both sides to a single flux. In this work, the local Lax-Friedrichs flux formulation is used,

$$g(U^+, U^-, n) = \frac{1}{2}\left(f(U^+) + f(U^-)\right) \cdot n + \frac{1}{2}\left(\max_{U \in [U^+, U^-]} |a(U) \cdot n|(U^+ - U^-)\right), \tag{16}$$

where a is the vector of eigenvalues of the flux Jacobian. The solution U is approximated by a polynomial basis

$$U(\boldsymbol{\xi}, t) \approx \sum_{i,j,k=0}^N \bar{u}_{ijk}(t)\psi_{ijk}(\boldsymbol{\xi}), \qquad \psi_{ijk}(\boldsymbol{\xi}) = l_i(\xi^1)l_j(\xi^2)l_k(\xi^3), \tag{17}$$

Fig. 3 Legendre-Gauss
nodes in a 2D reference
element for $N = 3$

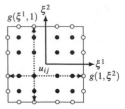

where the basis functions ψ_{ijk} are the product of one-dimensional Lagrange polynomials l of degree N in each spatial direction and $\bar{u}_{ijk}(t)$ are the coefficients to be
determined. The nodal basis is defined on a set of interpolation points $\{\xi\}_{i=0}^{N}$ on the
interval $\xi \in [-1, 1]$, which in this work are the Legendre-Gauss nodes (Fig. 3). For
the fluxes, the same approach is used for the approximation.

The three integrals in Eq. (15) are approximated by Gauss quadrature. Generally,
the Gauss quadrature of an arbitrary function $f(x)$ on the interval $[a, b]$ with $N + 1$
nodes can be written as

$$\int_{a}^{b} f(x)\,\mathrm{d}x \approx \sum_{i=1}^{N} \omega_i f(x_i), \tag{18}$$

where the weights ω_i and the integration nodes x_i are specific to the chosen quadrature
type. These weights are pre-calculated and stored to make the algorithm efficient.
With the interpolation points $\{\xi_i\}$ collocated at the Gauss nodes, all sums collapse
into single values, yielding the discrete DG operator $\mathscr{L}(U, t) = U_t$ [17]. In the next
step, the semi-discrete formulation is integrated in time to obtain the solution at the
next time step, for which a low-storage fourth-order Runge-Kutta scheme is used [5].

3.3 Finite-Volume Method for the Flow Simulation

A second-order finite-volume method is used to solve the unsteady Navier-Stokes
equations for compressible flow as given in Sect. 2.1. The solver has been extensively validated and used for various flow problems previously [15, 16]. A detailed
description of the method can be found in [13, 15, 16, 28].

4 Coupling Strategy

To solve the acoustic perturbation equations, the averaged quantities $\bar{u}$ and $\bar{c}$ and
the source term q_m have to be determined first. The flow solution is advanced without coupling until the averaged quantities are statistically converged. The coupling
process for each time step of the LES reads:

1. Advance the LES solution.
2. Calculate the source terms from instantaneous and averaged quantities.
3. Advance the CAA solution.

The actual coupling takes place via the source terms computed from the LES solution, which are then used to solve the APE. This means that there is a one-way coupling from the flow solution to the acoustic field, while the flow solution is not influenced by the acoustic field.

In the direct-hybrid method described here, the LES and the CAA simulation are both performed within a single simulation framework and by using the same grid topology. This makes certain aspects of the coupling process more efficient and allows a more fine-grained control over the interface between the two solvers. In the following, some details of the method are presented.

4.1 Spatial Coupling

The instantaneous variables of the source term q_m are available after each time step from the flow simulation. They have to be transferred, however, from the LES to the acoustic grid. Since both simulations typically operate on different levels of the same grid, identification of corresponding cells is possible by traversing the octree constituting the hierarchical Cartesian mesh. While LES and CAA leaf cells can generally be of different size, the coupling always happens within a single subtree. Since the domain decomposition algorithm distributes entire subtrees on different processes (see also Sect. 3.1), no additional inter-rank communication is required for the exchange of data between CFD and CAA cells.

This type of mesh also guarantees that there are no partially overlapping cells, i.e., a smaller cell is always fully contained inside a larger cell. Note that the DG elements are generally of higher order than the finite-volume cells. Depending on the resolution of the fluid and acoustics problems, four types of transformations are possible.

In the simplest case, one fluid cell corresponds exactly to one acoustics cell (Fig. 4a). That is, the source term is calculated once in the finite-volume part and the same value is used at all Gauss nodes of the DG element. This approach is used exclusively in the present work, i.e., no spatial interpolation is performed. Similar to the one-to-one mapping, the source term is calculated once and then used at all Gauss nodes of all elements if one fluid cell is mapped to multiple acoustics cells (Fig. 4b).

Having multiple finite-volume cells mapped onto one DG element (Fig. 4c) requires the values at the Gauss nodes to be interpolated from several flow cells. A natural choice would be to interpret the finite-volume cells as equidistant nodes of a polynomial and to obtain the values at the Gauss nodes through projection. This, however, can lead to spurious oscillations if the number of finite-volume cells and thus the polynomial degree is high, especially in regions with large flow gradients.

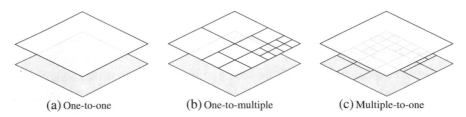

(a) One-to-one   (b) One-to-multiple   (c) Multiple-to-one

Fig. 4 Possible spatial mappings for coupled simulations. Aeroacoustics cells (*top*) are *white*, fluid cells (*bottom*) are *grey*

Other possibilities are weighted least squares methods, nearest neighbor interpolation, or inverse distance weighting. Which approach is best depends on a number of factors. A practical consideration is the computational cost of the chosen method, e.g., whether the effort scales linearly with the number of degrees of freedom or worse, since the interpolation has to take place at each flow simulation time step. The smoothness of the interpolated function is also important, especially in high-gradient zones. Furthermore, it is desireable to have a conservative interpolation scheme such as proposed by Farrell and Maddison [8], to avoid distorting the source terms.

If there are regions without either a flow or acoustics grid, no coupling is performed. If only acoustic cells exist, far-field values for the averaged quantities $\bar{c}$ and $\bar{u}$ have to be specified for the APE, e.g., the freestream values from the flow field. The source term q_m is set to zero with a smooth transition from non-zero to zero values.

4.2 Temporal Coupling

The coupling between the flow and the acoustics simulations has to be realized at each time step. Due to the explicit global time stepping it is possible that the time step size differs between the two solvers. In this case, at each time step the source term from the LES solution needs to be interpolated to the simulation time of the CAA solver.

Depending on the features of the geometry, the time step for the aeroacoustics simulation may be smaller than that for the flow simulation or vice versa and thus the source terms have to be interpolated between two flow time steps. As for the spatial coupling, there are many different interpolation methods to choose from. Linear interpolation is the most straightforward approach, with sometimes inferior results. Several temporal interpolation methods suitable for hybrid aeroacoustics simulations are compared and evaluated by Geiser et al. [10] and least-squares optimized interpolators were found to have the best properties when it comes to broadband error reduction.

The simplest approach is using the same time step for both simulations, which requires no interpolation between the two datasets. In this case, the next time step based on the CFL condition is determined for the CFD and the CAA method and the minimum of both methods is used, which is also the procedure that is used in this work.

4.3 Data Transfer

There exist two options for transferring data between the flow solution and the acoustics solution: via data files written to disk, i.e., offline coupling as used in standard hybrid approaches, or through in-memory data access, i.e., online coupling as done in the new direct-hybrid approach. Both methods are discussed in the following.

In offline coupling, the processes of obtaining the flow solution and running the aeroacoustics simulation are completely separated. At first, the flow solution is obtained and the source term q_m is written to a file at certain time intervals. During the acoustics simulation, the source terms are determined from the files by interpolation in time. Conceptually, this is the simplest approach, since except for the I/O routines nothing has to be changed inside the two simulations. However, the high amount of data that has to be transferred to and from the disk makes this method expensive in terms of computational cost, especially for large-scale simulations on thousands of cores. However, it is also the first step towards a simulation which makes use of online coupling as outlined next.

In online coupling, the flow and the acoustics simulations are fully integrated and run synchroneously at the same time. Typically, the flow solution will be advanced by one time step and the acoustics solution has to be updated until they are both synchronized. Since no files have to be written to disk, this approach is more efficient than offline coupling. If the acoustics cells are kept on the same computational core as the corresponding flow cells, the acoustics simulation can directly access the relevant information by simple memory transfer operations. This locality of data is achieved by the specific subdomain decomposition, which operates on the joint LES-CAA grid. On the other hand, the increased memory consumption makes it necessary to use more computational cores. Furthermore, due to the different number of operations for the finite-volume and the DG operator, paired with different numbers of flow cells per acoustics cell, load balancing between the cores becomes mandatory to achieve reasonable parallel efficiency. This is accomplished by assigning appropriate loads to the fluid and acoustics cells.

5 Results

The CFD solver has already been extensively tested and used in the past, e.g., in [13, 15, 16, 23, 28]. Thus in Sect. 5.1, only the new CAA solver is validated. Additionally, parallel performance results for the CAA solver are presented in Sect. 5.2.

5.1 *Validation of the Aeroacoustics Solver*

The DG method described in Sect. 3.2 is validated by solving the acoustic perturbation equations for several generic problems. It is demonstrated that the solver is able to correctly predict the acoustic pressure field for sheared mean flow, for acoustic reflection at a solid wall, and for sound waves emanating from a boundary layer.

5.1.1 Monopole in Sheared Mean Flow

Figure 5 shows the results for wave propagation in a sheared mean flow. The example was chosen since mixing layer-type flow configurations with sheared mean flow are typical for noise generation, e.g., for turbulent jets. An S-shaped velocity profile is prescribed for the mean velocity,

$$\bar{u} = \frac{1}{2} \tanh \left(\frac{2y}{\delta_w} \right), \tag{19}$$

where the shear-layer thickness is set to $\delta_w = 50$ and an analytical source term is used to generate an acoustic monopole [6]. The domain was discretized using 200×200 elements with a polynomial degree $N = 3$. Figure 5 shows the result in comparison to the perturbed pressure field obtained in [6] from the linearized Euler equations (LEE). It can be seen that the DG results agree well with the reference solution.

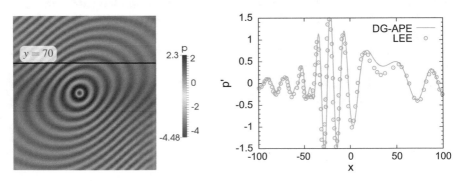

Fig. 5 Monopole in sheared mean flow (*left* perturbed pressure p', *right* p' at $y = 70$ and $t = 180$)

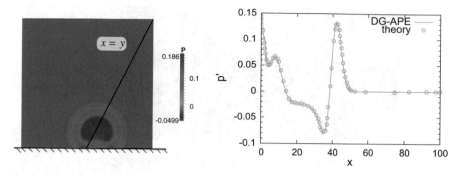

Fig. 6 Reflection of a pressure pulse at a solid wall (*left* perturbed pressure p', *right* p' at $x = y$ and $t = 30$)

5.1.2 Acoustic Reflection at a Solid Wall

A pressure pulse impinging on a plane wall in the presence of a uniform mean flow was simulated to validate the wall boundary conditions. The wall is located at $y = 0$ and the initial conditions at time $t = 0$ are

$$u' = v' = 0, \qquad p' = \exp\left\{-(\ln 2)\frac{x^2 + (y - 25)^2}{25}\right\}. \tag{20}$$

The mean flow is prescribed parallel to the wall by setting $\bar{u} = 0.5$, $\bar{v} = 0.0$. Both the setup and the analytical values are taken from [14]. The square-shaped computational domain with side length $l = 200$ was discretized using 256 elements in each spatial direction with a polynomial degree of $N = 5$. In Fig. 6, results for the acoustic pressure field of the reflected pulse are shown. They confirm that the CAA solver is able to correctly predict the reflection of acoustic waves from a solid wall.

5.1.3 Monopole in a Boundary Layer

In this case, a plane sound wave is assumed to travel through a small channel and to exit through a small orifice in a plane wall. Due to the small size of the channel, the emanating wave is an approximation for a singular monopole at the wall [3]. The domain is defined by $x \in [-25.6, 25.6]$ in the x-direction and $y \in [0.0, 20.0]$ in the y-direction, and it is discretized by 400,000 elements with a polynomial degree of $N = 3$. The monopole has a size of $\epsilon = 0.1$ and is located at the origin. It is created by enforcing a sinusoidal boundary state by setting

$$u' = 0, \qquad v' = p' = \sin(2\pi t). \tag{21}$$

Fig. 7 Contour plot of perturbed pressure p' for monopole in boundary layer

Fig. 8 Directivities for rms
pressure along $r = 15$

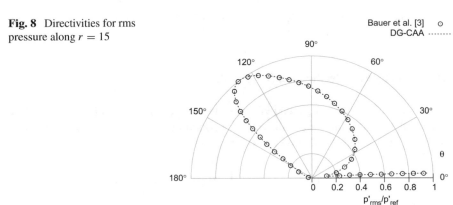

In addition to the monopole at the wall, a non-zero mean velocity is prescribed, which
decreases to zero in the boundary layer region:

$$\bar{u} = \begin{cases} M_x(2y - 2y^2 + y^4), & \text{if } 0 \leq y \leq 1, \\ M_x, & \text{if } y > 1, \end{cases} \quad \bar{v} = 0, \quad (22)$$

where the Mach number is set to $M_x = 0.3$. Figure 7 shows a contour plot of the resulting pressure field. In Fig. 8 the results are compared to those in [3]. The DG-CAA
solution is virtually indistinguishable from the reference solution, which demonstrates that the DG-APE method is able to adequately capture the refraction and
reflection of sound waves in flow fields with velocity gradients, both in the channel
region at $\theta < 10°$ and in the shadow region at $140° < \theta < 180°$.

5.2 Parallel Performance Analysis

To assess the parallel performance of the newly developed aeroacoustics solver, a
strong scaling experiment with two setups was performed on HPC systems. In each

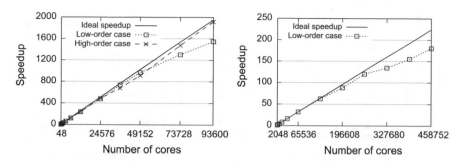

Fig. 9 Strong scaling experiments for the CAA solver on a Cray XC 40 (*left*) and a BlueGene/Q (*right*)

setup, the three-dimensional domain is cube-shaped. To obtain meaningful error measures, a manufactured solution approach was used, i.e., an auxiliary source term was added to the system of equations such that the analytical initial conditions, which are based on trigonometric functions, fulfill the system of equations exactly. In the first setup, a grid with 16.8 million cells and a polynomial degree $N = 3$ was used (low-order case). For the second setup, the number of cells was reduced to 2.1 million and the polynomial degree was set to $N = 7$ (high-order case). This yields the same global number of degrees of freedom for both cases (1.1 billion). The setups were chosen to be representative of typical large-scale aeroacoustics simulations under realistic conditions.

Figure 9 shows the strong scaling results for both setups on two state-of-the-art supercomputers, i.e., the Cray XC 40 of the High-Performance Computing Center Stuttgart and the BlueGene/Q of the Forschungszentrum Jülich. On both machines, the simulations were executed with one MPI rank per core and two OpenMP threads per rank. For the Cray system, the low-order case has a parallel efficiency of 79 % on 93,600 cores, which improves to 98 % for the high-order case. Both values are very satisfactory. On the BlueGene/Q, the efficiency for the low-order case on the full machine is 80 %. From these results, it can be concluded that the CAA solver is highly scalable and that it is well-suited for large-scale aeroacoustics simulations. Furthermore, the comparison of the two setups on the Cray XC 40 shows that it is beneficial for the parallel efficiency to use a higher-order approximation in the DG scheme.

To highlight the necessity of developing a new coupling approach for hybrid CFD-CAA simulations, another scaling experiment was conducted. In this case, a CAA simulation of a two-dimensional mixing layer was performed with offline coupling, i.e., the source term information was read from data files [27]. Figure 10 shows the speedup and the absolute wall-clock time for a single-threaded scaling from 32 to 4,096 cores. In the left figure, the speedup is shown once for the overall simulation, with an ultimate efficiency of 61 % at 4,096 cores. When excluding the I/O time, i.e., the time spent reading the source term data from disk, the efficiency improves to 92 %. The reason for this behavior can be understood when looking at the

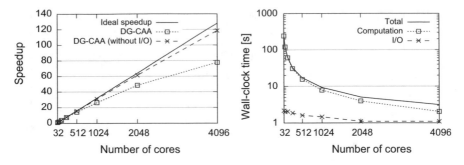

Fig. 10 Speedup (*left*) and wall-clock time (*right*) for a offline coupling simulation with 31.4 million cells and $N = 1$ on a BlueGene/Q

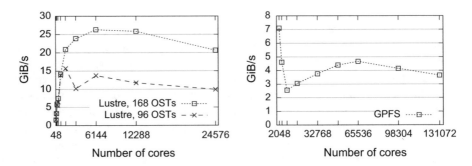

Fig. 11 Strong scaling results for the I/O write performance of a 63 GiB file on a Cray XC 40 using two Lustre file systems with 96 and 168 object storage targets (OST) respectively (*left*) and a BlueGene/Q using a GPFS file system (*right*)

wall-clock time for computation and I/O separately (see right figure): while the time for computation continuously decreases when using higher core counts, the curve for I/O time flattens out when going from 2,048 to 4,096 cores. This means that the I/O component ceases to scale beyond a certain number of cores, effectively turning the I/O into a bottleneck for the overall simulation.

The degradation of the parallel efficiency for offline-coupled simulations due to I/O performance limits can be further substantiated by examining the I/O bandwidth on current HPC systems. In Fig. 11, the measured maximum write speed for a single 63 GiB file is shown at increasing numbers of cores. The numbers were obtained with the Parallel netCDF library [20] using collective I/O and one MPI rank per core. On both machines, i.e., a Cray XC 40 with a Lustre file system (left figure) and a BlueGene/Q with a GPFS file system (right figure), the I/O bandwidth peaks at a certain number of cores and actually decreases for higher core counts. These results strongly suggest the need for an online coupling approach, where the CFD and the CAA solvers do not have to rely on the file I/O system to exchange data.

6 Conclusions

A direct-hybrid method suitable for large-scale aeroacoustic simulations has been presented. The flow field is predicted using an LES solver based on the finite-volume method. For the CAA solution, a nodal DG method is used to solve the acoustic perturbation equations for the determination of the acoustic field. In the novel approach, both solvers use the same hierarchical Cartesian grid, enabling an efficient data exchange between the two solvers. Appropriate strategies for the spatial and temporal coupling are described.

The CAA method is shown to correctly predict the acoustic pressure field for a monopole in sheared mean flow, acoustic reflection at a solid wall, and a monopole in a boundary layer. In addition, the parallel performance of the new scheme is investigated in several strong scaling experiments. They show that the new DG-CAA solver is capable of efficiently running simulations on hundreds of thousands of cores. Furthermore, while the direct-hybrid method with offline coupling involving disk I/O scales well up to a 128-fold increase in MPI ranks, the I/O operations necessary for reading the source terms from disk are identified as a bottleneck towards extreme scaling. This observation is further corroborated by an analysis of the I/O bandwidth on two current HPC systems, which emphasizes the need for the online coupling approach.

Overall, the proposed direct-hybrid method has shown to be a good candidate for efficient, highly parallel CAA simulations. As a next step, spatial as well as temporal interpolation schemes need to be investigated to lessen the restriction on the resolution requirements in space and time. A dynamic load balancing scheme will be developed to further improve the parallel performance for moving geometries.

Acknowledgements This work has been performed with the support from the JARA-HPC SimLab Fluids & Solids Engineering of the RWTH Aachen University, Germany and the Forschungszentrum Jülich, Germany. The authors gratefully acknowledge the allocation of supercomputing time as well as the technical support by the High-Performance Computing Center Stuttgart of the University of Stuttgart, Germany and by the Jülich Supercomputing Centre of the Forschungszentrum Jülich, Germany.

References

1. Atkins, H.L.: Continued development of the discontinuous Galerkin method for computational aeroacoustic applications. AIAA Paper (97-1581) (1997)
2. Bassi, F., Rebay, S.: A high-order accurate discontinuous finite element method for the numerical solution of the compressible Navier-stokes equations. J. Comput. Phys. **131**(2), 267–279 (1997)
3. Bauer, M., Dierke, J., Ewert, R.: Application of a discontinuous Galerkin method to discretize acoustic perturbation equations. AIAA J. **49**(5), 898–908 (2011)
4. Bui, T.Ph., Schröder, W., Meinke, M.: Numerical analysis of the acoustic field of reacting flows via acoustic perturbation equations. Comput. Fluids **37**(9), 1157–1169 (2008)

5. Carpenter, M.H., Kennedy, C.: Fourth-order 2N-storage Runge-Kutta schemes. NASA Report TM 109112, NASA Langley Research Center (1994)
6. Ewert, R., Schröder, W.: Acoustic perturbation equations based on flow decomposition via source filtering. J. Comput. Phys. **188**, 365–398 (2003)
7. Ewert, R., Schröder, W.: On the simulation of trailing edge noise with a hybrid LES/APE method. J. Sound Vibr. **270**(3), 509–524 (2004)
8. Farrell, P., Maddison, J.: Conservative interpolation between volume meshes by local Galerkin projection. Comput. Meth. Appl. Mech. Eng. **200**(1–4), 89–100 (2011)
9. Flad, D., Frank, H., Beck, A.D., Munz, C.D.: A Discontinuous Galerkin spectral element method for the direct numerical simulation of aeroacoustics. AIAA Paper (2014-2740) (2014)
10. Geiser, G., Marinc, D., Schröder, W.: Comparison of source reconstruction methods for hybrid aeroacoustic predictions. International Journal of Aeroacoustics **12**(7–8), 639–662 (2014)
11. Geiser, G., Schlimpert, S., Schröder, W.: Thermoacoustical noise induced by laminar flame annihilation at varying flame thicknesses. In: 18th AIAA/CEAS Aeroacoustics Conference (33rd AIAA Aeroacoustics Conference), 04–06 June 2012, Colorado Springs, CO, AIAA 2012–2093 (2012)
12. Gröschel, E., Schröder, W., Renze, P., Meinke, M., Comte, P.: Noise prediction for a turbulent jet using different hybrid methods. Comput. Fluids **37**(4), 414–426 (2008)
13. Günther, C., Meinke, M., Schröder, W.: A flexible level-set approach for tracking multiple interacting interfaces in embedded boundary methods. Comput. & Fluids **102**, 182–202 (2014)
14. Hardin, J., Ristorcelli, J.R., Tam, C.K.W. (eds.): ISCASE/LaRC Workshop on Benchmark Problems in Computational Aeroacoustics (CAA), vol. NASA Conference Publication 3000. NASA (1995)
15. Hartmann, D., Meinke, M., Schröder, W.: An adaptive multilevel multigrid formulation for Cartesian hierarchical grid methods. Comput. Fluids **37**, 1103–1125 (2008)
16. Hartmann, D., Meinke, M., Schröder, W.: A strictly conservative Cartesian cut-cell method for compressible viscous flows on adaptive grids. Comput. Meth. Appl. Mech. Eng. **200**, 1038–1052 (2011)
17. Hindenlang, F., Gassner, G.J., Altmann, C., Beck, A., Staudenmaier, M., Munz, C.D.: Explicit discontinuous Galerkin methods for unsteady problems. Comput. Fluids **61**, 86–93 (2012)
18. Koh, S., Schröder, W., Meinke, M.: Turbulence and heat excited noise sources in single and coaxial jets. J. Sound Vibr. **329**, 786–803 (2010)
19. Kopriva, D., Woodruff, S., Hussaini, M.: Discontinuous spectral element approximation of Maxwell's equations. In: B. Cockburn, G. Kariadakis, C.W. Shu (eds.) Proceedings of the International Symposium on Discontinuous Galerkin Methods. Springer (2000)
20. Li, J., Zingale, M., Liao, W.k., Choudhary, A., Ross, R., Thakur, R., Gropp, W., Latham, R., Siegel, A., Gallagher, B.: Parallel netCDF: a high-performance scientific I/O interface. In: Proceedings of the 2003 ACM/IEEE Conference on Supercomputing - SC '03, p. 39. ACM Press, New York, USA (2003)
21. Lintermann, A., Schlimpert, S., Grimmen, J.H., Günther, C., Meinke, M., Schröder, W.: Massively parallel grid generation on HPC systems. Comput. Meth. Appl. Mech. Eng. **277**, 131–153 (2014)
22. Liu, J.G., Shu, C.W.: A high-order discontinuous Galerkin method for 2D incompressible flows. J. Comput. Phys. **160**(2), 577–596 (2000)
23. Pogorelov, A., Meinke, M., Schröder, W.: Cut-cell method based large-eddy simulation of tip-leakage flow. Phys. Fluids **27**(7), 075106 (2015)
24. Reed, W., Hill, T.: Triangular mesh methods for the neutron transport equation. Tech. Rep. LA-UR-73-479, Los Alamos Scientific Laboratory (1973)
25. Directorate-General for Research, Innovation European Union: Flightpath 2050: Europe's Vision for Aviation: Maintaining Global Leadership and Serving Society's Needs. Office for Official Publications of the European Communities (2011)
26. Sagan, H.: Space-filling curves, 1st edn. In: Universitext. Springer, New York (1994)
27. Schlottke, M., Cheng, H.J., Lintermann, A., Meinke, M., Schröder, W.: A direct-hybrid method for computational aeroacoustics. In: AIAA Aviation, 22–26 June 2015, Dallas, TX, 21st AIAA/CEAS Aeroacoustics Conference, AIAA-2015-3133 (2015)

28. Schneiders, L., Hartmann, D., Meinke, M., Schröder, W.: An accurate moving boundary formulation in cut-cell methods. J. Comput. Phys. **235**, 786–809 (2013)
29. Yakovlev, S., Xu, L., Li, F.: Locally divergence-free central discontinuous Galerkin methods for ideal MHD equations. J. Comput. Sci. **4**(1–2), 80–91 (2013)

HPC Applications for Manufacturing Innovation in Aerospace Fields

Ryoji Takaki and Seiji Tsutsumi

Abstract JAXA promotes research and development of High Performance Computing technology in order to help aerospace developments. From April, 2016, JAXA started full operation of JAXA Supercomputer System 2:JSS2. A main engine of JSS2 is called SORA-MA, which is FUJITSU Supercomputer PRIMEHPC FX100. It is a many-core based scalable parallel cluster system. With supercomputer systems, JAXA has been driving the incorporation of numerical simulation technologies into the design process in order to innovate the manufacturing in aerospace fields. This paper reports brief overview and preliminary performance evaluation of JAXA's new supercomputer system. An example of numerical simulations applied to a development of a launch-pad for a new rocket called Epsilon is also presented.

1 Introduction

Japan Aerospace Exploration Agency (JAXA) is a national aerospace agency in Japan. JAXA was born in 1 October 2003, through the merger of three previously independent organizations. JAXA is a core agency to support the Japanese government's overall aerospace development and utilization. Therefore, JAXA has been working on a variety of aerospace activities: space and planetary science research by asteroid probe HAYABUSA, planetary probes and astronomy satellites, development of space transportation systems like H-IIA, H-IIB and Epsilon launch vehicles, human space activities such as Japanese module of International Space Station, astronauts and unmanned cargo transporter HTV, utilization with satellite for earth observation, communication and navigation using various satellite, aviation programs for the next generation airplane and jet engines. Basic technology research and educations are also important missions in JAXA.

R. Takaki (✉) · S. Tsutsumi
Japan Aerospace Exploration Agency, 3-1-1 Yoshinodai, Chuo-ku,
Sagamihara, Kanagawa 252-5210, Japan
e-mail: ryo@isas.jaxa.jp

S. Tsutsumi
e-mail: tsutsumi.seiji@jaxa.jp

© Springer International Publishing AG 2016
M.M. Resch et al. (eds.), *Sustained Simulation Performance 2016*,
DOI 10.1007/978-3-319-46735-1_13

159

One of the HPC (High Performance Computing) center in JAXA is JAXA's Engineering Digital Innovation Center, called JEDI Center. The name of the center "JEDI" was determined by the belief that JAXA's simulation and digital engineering technology will be a strong help for the credibility improvement of JAXA's activity. The objective of this center is a contribution to space development by applying information technologies and simulation technologies. Three key roles are defined as the main task of JEDI center to realize the previous objective. First is an introduction of up-to-date IT (Information Technology) into the spacecraft and aircraft projects. Second is a research and development on numerical simulation technology and its application to the spacecraft and aircraft projects. The last is introduction and operation of JAXA's supercomputer system and research and development of its application technology. Currently, the name of JEDI was formally disappeared due to an organizational change of JAXA. However, the spirit of JEDI center has been preserved and the our group is informally still called "JEDI center".

This paper reports brief overview and preliminary performance evaluation of JAXA's new supercomputer system. An example of numerical simulations applied to a development of a launch-pad for a new rocket called Epsilon is also presented.

2   JAXA's Supercomputer System 2

2.1   System Overview

From April 2016, JAXA started full operation of JAXA's Supercomputer System 2 (JSS2), which consists of SORA (Supercomputer for earth Observation, Rockets and Aeronautics) and J-SPACE (JAXA's Storage Platform for Archiving, Computing and Exploring). Figure 1 shows system configuration of JSS2, which consists of several systems.

SORA-MA is a main computational engine. SORA-PP is a pre-post processing system, which is used for grid generation, visualization and data analysis of computed results. It is also used as a platform for commercial softwares because it has commonly used Intel CPUs. SORA-LM has a large memory and it is used for non-parallelized applications which require large memory space. SORA-LI is a login system and SORA-FS is a file system for JSS2. SORA-TPP and SORA-TFS are a local sever and a local file system at Tsukuba region. SORA-KFS and SORA-SFS are local file systems at Kakuda region and Sagamihara region, respectively. These systems support remote users to use SORA-MA and SORA-PP through the network from Tsukuba, Kakuda and Sagamihara.

SORA-MA is FUJITSU Supercomputer PRIMEHPC FX100 [1] (cf. Fig. 2). It is a many-core based scalable parallel cluster system, which has 3,240 compute nodes. Each compute node has one CPU:SPARC64^{TM} XIfx chip of 2.2 GHz clock speed and is connected by Tofu2 (Torus Fusion 2) interconnect. HMC (Hybrid Memory Cube) is used as a main memory of nodes, which shows 480 GB/s memory bandwidth:

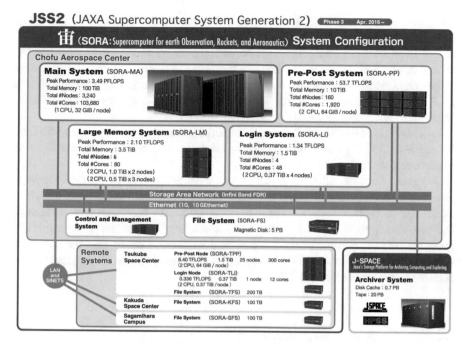

Fig. 1 System configuration of JSS2

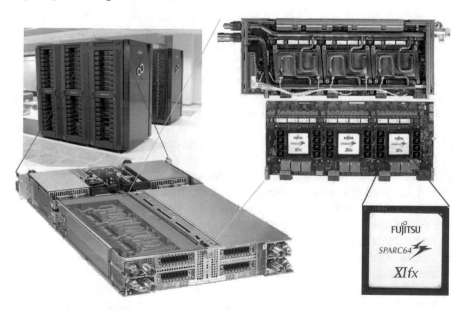

Fig. 2 SORA-MA (Fujitsu Supercomputer PRIMEHPC FX100)

240 GB/s for read and 240 GB/s for write. The bandwidth of Tofu2 is 12.5 GB/s. SPARC64^{TM}XIfx has 32 cores and 2 assistant cores. The theoretical peak performance is about 1TFLOPS. Assistant cores are used by system daemons, file I/O processes and data transfer processes, which help performance advances in large scale computations. Thirty-two cores (and also 2 assistant cores) are divided into 2 CMGs (Core Memory Groups), which shares L2 Cache. Each core has 8 FMAs (Floating-point Multiply and Add) and 4-wide SIMD.

SORA-PP is FUJITSU PRIMERGY RX350 S8. Each compute node has two Intel Xeon E5-2643V2 CPUs of 3.5 GHz. There are 6 cores on one CPU, hence 12 cores in one compute node.

Because of an internal hierarchical structure (core/CMG/CPU/node) of the compute node, several parallel programming models, a Flat-MPI model and a Hybrid model can be applicable. In the Flat-MPI model, 32 MPI processes can run at each compute node. In the Hybrid model, several combinations of the number of MPI processes and the number of threads can be considered. Considering the hardware structure, a combination of 2 MPI processes and 16 threads are recommended, where each MPI process run on each CMG in CPU.

2.2 Performance Measurements

The performance of SORA-MA was measured by the basic benchmark program STRAM [2] and an application program UPACS-Lite.

Figure 3 shows results of STREAM TRIAD on SORA-MA. In this test, one compute node with 16 cores was used; one CMG was used by using the Hybrid model with one MPI process and 16 threads. There are 3 lines in this figure, showing same performance with different type of Fortran arrays: a static array, an allocatable array and a pointer array. The theoretical peak memory bandwidth is 180 GB/s on one CMG (namely 16 cores). Therefore the efficiency is about 85.7 %. Same efficiency is also obtained by the case using one CPU with 32 cores.

UPACS-Lite was used to measure the performance of SORA-MA. UPACS-Lite is a full kernel of a real application program UPACS [3], which has been used to conduct several computations in order to support JAXA's projects. UPACS-Lite is a typical compressible Navire-Stokes solver. It is a stencil type program. Features of UPACS-Lite are as follows, multi-block structured grid method, implicit time integration with 2nd order inner iterations and hybrid parallel model (MPI and OpenMP are used). Block Red-Black parallelization are applied to the time integration for parallelization. In this multi-block structured grid method, a computational space is divided into several blocks and each block is represented by structured grid. Physical values such as density, velocity, pressure and so on, are assigned to each structured grid point in every blocks. This physical structure is mapped to the data structure of the program. Moreover, each block must have different size and shape because of the convenience of grid generation. Considering these features of multi-block structured grid, Fortran functions of structure and dynamic allocatable array in Fortran 90 are applied. As for

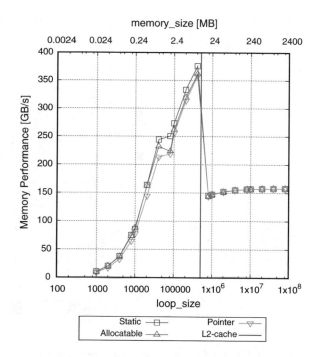

Fig. 3 Memory access performance of SORA-MA (FX100) measured by STREAM TRIAD

parallelization, conventional domain decomposition method is used. Several blocks are assigned to each MPI process. This feature is vary important because it can give more flexibility to create multi-block structured grids.

Figure 4 shows the comparison of computational time among FX1, FX10, FX100 (SORA-MA) and Intel (SORA-PP), where one compute node was used and program tunings for FX10 and FX100 were applied. FX1[1] and FX10 are Fujitsu's supercomputers of preceding generation of FX100.

Red bars show computational time for certain test condition. Blue line shows speedup ratio to FX1 and green line shows efficiency related to FX1, where the efficiency is calculated as the speedup ratio divided by the peak performance ratio to FX1. In this figure, "$(aPbT)$" represents the core usage on the compute node; a is the number of processes and b is the number of threads. Regarding FX100, the case of 16 MPI processes with 2 threads is the fastest, showing 37 times faster than FX1, where the peak performance ratio of the hardware between FX1 and FX100 is 25. The case of typical Hybrid model (2 processes with 16 threads), which fits the hardware structure shows 32 times faster than FX1. The Flat-MPI model is often faster than any Hybrid models currently because the computational scale of most computations in JAXA was not so big. As computational scales become larger then the superiority of the Hybrid model comes to be observed.

---

[1]FX1 was a main computational engine of previous system:JSS1. Peak performance of the FX1 node is 40GFLOPS.

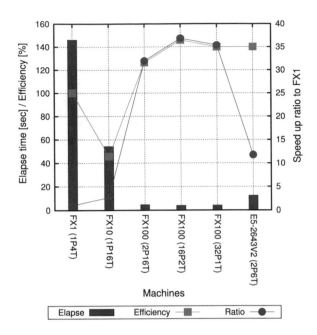

Fig. 4 Comparison of computational time of UPACS-Lite with software tuning among FX1, FX10, FX100 and Intel

As for FX100, following program tunings were applied. Regarding data structure, array of structure (AOS) is changed to structure of array (SOA); index order of multi-dimensional arrays is optimized. Regarding loop structure, outer two loops are collapsed by the OpenMP directive to get better scalability for sixteen threads; innermost loops are unrolled to promote SIMD utilization. The order of multiple loops and indices of multi-dimensional arrays are optimized by trial and error, shown in Table 1. Here, indices i, j, k show positions in the computational space, whose max values $imax, jmax, kmax$ are $O(10) \sim O(100)$. The index n indicates physical values whose max value $nPhys$ is $O(1)$.

Figure 5 shows the comparison of computational time of UPACS-Lite without software tuning, which means the comparison of hardware capabilities. In this case, Flat-MPI model is the fastest, showing 16 time faster than FX1. Green line shows the decrease of the efficiency of FX10 and FX100 compared to FX1, which means the speedup tuning is indispensable to utilize the hardware features.

3 Application to the Project

A example of HPC application to JAXA's project is presented here. A new rocket called Epsilon [4] has been developed in JAXA and 1st Epsilon rocket was launched in Oct. 2013. Epsilon rocket is a solid propellant rocket, whose aims are to reduce the cost by a third of that for the former M-V rocket as well as to lower hurdles to space by making rocket launches much simpler.

Table 1 Effects of the index order of multiple loops and multi-dimensional arrays

Source program	Time (s)	Source program	Time (s)
do n = 1, nPhys !$omp parallel do do k = 1, kmax; do j = 1, jmax do i = 1, imax a(i, j, k, n) = ... enddo enddo; enddo !$omp end parallel do enddo	0.320	do n = 1, nPhys !$omp parallel do do k = 1, kmax; do j = 1, jmax do i = 1, imax a(n, i, j, k) = ... enddo enddo; enddo !$omp end parallel do enddo	0.703
!$omp parallel do do k = 1, kmax; do j = 1, jmax do i = 1, imax a(i, j, k, 1) = ... a(i, j, k, 2) = ... a(i, j, k, nPhys) = ... enddo enddo; enddo !$omp end parallel do	0.247	!$omp parallel do do k = 1, kmax; do j = 1, jmax do i = 1, imax a(1, i, j, k) = ... a(2, i, j, k) = ... a(nPhys, i, j, k) = ... enddo enddo; enddo !$omp end parallel do	0.246

!$omp parallel do
do k = 1, kmax; do j = 1, jmax
do n = 1, nPhys
do i = 1, imax
a(i, j, k, n) = ...
enddo;
enddo
enddo; enddo
!$omp end parallel do
	0.250	!$omp parallel do
		do k = 1, kmax; do j = 1, jmax
		do n = 1, nPhys
		do i = 1, imax
		a(n, i, j, k) = ...
		enddo;
		enddo
		enddo; enddo
		!$omp end parallel do
			0.268

There are several subjects in the rocket development. Acoustic loading generated by the rocket propulsion system is one of such critical issues. Low-frequency oscillation by the plume acoustics is observed during the launch of the rockets in the world. Acoustic power, radiated from the Epsilon rocket is estimated to be about 3 MW. It is extremely huge comparing a typical peak power of usual audio device, which is about 40 W even at full volume operations. This low-frequency oscillation may cause critical troubles in rockets and spacecraft on board. Therefore, it is necessary to accurately predict and ease the acoustic environment. To improve acoustic environment, various methods are applied to the launch-pad in the world. Each launch-pad has it's own concept and ideas to reduce acoustic environment. It is a problem how to develop the launch-pad for the newly developed Epsilon rocket.

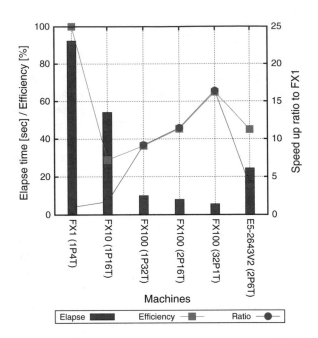

Fig. 5 Comparison of computational time of UPACS-Lite without software tuning among FX1, FX10, FX100 and Intel

Prediction of the acoustic loading on spacecraft is not established well. Currently the empirical method proposed in NASA SP-8072 [5] is the only way to predict the acoustic loading. However, this method doesn't have enough reliability, accuracy and applicability because it is not based on the physical basis but on the experimental data. Therefore, it is not sure if this method can be applied to JAXA's new rocket and launch-pad. In fact, this empirical method, applied to the design of new Epsilon's launch-pad, suggests that it is necessary to dig a 70 m depth groove as a gas duct in order to achieve the required acoustic environment. It is too expensive and is not acceptable for the project. Therefore a new design method is necessary. It should be based on the physical basis of the plume acoustics. Therefore, numerical simulation was conducted in order to understand the physical mechanism of acoustic wave generation and propagation.

Figure 6 shows a numerical simulation of a impinging jet [6], where the jet and the oblique plate represent a rocket plume and a flame deflector in a launch-pad, respectively. This figure shows pressure distributions by color contours whose stripe pattern represents acoustic waves. The jet impinges on the oblique plate generating several acoustic waves. The generation and propagation of the acoustic waves can be seen clearly from this result. Numerical simulations can help to understand physical mechanism of acoustic wave generation and propagation. Therefore, numerical simulations can help to take effective action against the problems by not a heuristic approach but a rational approach. After understanding the physical mechanisms by numerical simulations, several concepts and ideas to reduce plume acoustics effec-

Fig. 6 Numerical simulation of a impinging jet. Color contours show pressure distributions

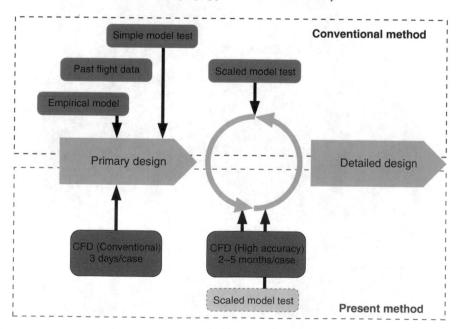

Fig. 7 A new design method by using numerical simulations (CFD)

tively were discussed. Moreover, a parametric study to optimize design parameters were conducted.

Figure 7 shows the design process applied to the Epsilon's launch-pad design.

In conventional method, empirical models, past flight data and simple model test are used at a primary design phase and several scaled model tests are conducted to ensure the design results at a detailed design phase. In our new method, numerical simulations using conventional CFD (Computational Fluid Dynamics) methods were conducted at the primary design phase. Conventional CFD can calculate in relatively shorter time (3 days per one case) and help to conduct many simulations as a parametric study. As the parametric study, 20 cases were conducted within 2 months by using 768 nodes (3, 072 cores) of JSS1 (Fujitsu FX1). After the primary design phase, precise but expensive simulations with a high accuracy method were applied at the detailed design phase. It took a few months for one case by using JSS1 at that time. Figure 8 shows an example of precise simulations conducted at the detailed design phase. There can be seen the rocket just after the lift-off, a launch-pad construction and acoustic waves around the launch-pad in this figure. A few scaled model tests were still used due to the limitation of CFD because CFD can't resolve

Fig. 8 Precise simulation of acoustic environment around the launch pad. Color contours show pressure distributions

higher frequency range of acoustics. The upper limit of frequency was 800 Hz at that time. New supercomputer JSS2 is expected to extend the upper limit of the frequency range up to 2,000 Hz, which is enough for the launch-pad design.

Thanks to knowledge and understanding of the physical mechanisms obtained from numerical simulations, several new features are adopted to realize low acoustic environment for Epsilon's launch-pad. Figure 9 shows Epsilon's launch-pad. The shape of the flame deflector inside the duct was optimized and concept of the lofty pad, exhaust duct on the ground and the roof shape of the duct are accomplished by numerical simulations. The cost to build this launch-pad designed by using numerical simulations, is reduced by tenth part compared to that designed by the conventional method.

Figure 10 shows the acoustic environment of several small and medium size rockets in the world. In this figure, a red bar shows Epsilon rocket, showing the world-leading quietness. Where, Pegasus XL seems to be the most quiet. However, it is an air-launched rocket and can't be compared with other ground launched rockets. This quietness is obtained not only by numerical simulations but also by the acoustic absorbent attached to the inside of the rocket fairing. However, numerical simulations played an important role in this launch-pad design.

In the development of Epsilon rocket, numerical simulations were appropriately utilized for the preliminary design and the detailed design of the launch-pad. Relatively light simulations using conventional methods were conducted in order to

Fig. 9 Epsilon's launch-pad

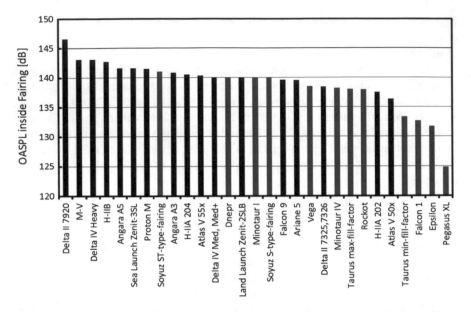

Fig. 10 Comparison of acoustic level for typical small and medium class rockets in the world

conduct parametric studies at the preliminary design phase and precise but expensive simulations were conducted at the detailed design phase to assess quantitatively. Numerical simulations can help to understand physical mechanisms and help to take effective action against the design problems by not a heuristic approach but a rational approach. This is an innovative design approach which is not a empirical or experimental based design but a simulation based design. In this launch-pad development, the simulation based design contributed to the cost reduction to one-tenth as well as to the reduction of acoustic level to one-tenth, which is the top level performance in the world.

4 Conclusions

Overview of JAXA's new supercomputer system JSS2 is presented. Main computational system is SORA-MA, which is a many-core based scalable parallel cluster system. The performance of SORA-MA is evaluated by the basic benchmark program STRAM and the application program UPACS-Lite. The result obtained by UPACS-Lite shows the importance of program tuning in order to utilize the hardware features.

The HPC application applied to the aerospace development in JAXA is also presented. New design method using HPC technology is applied to the launch-pad design of newly developed Epsilon rocket. Thanks to this innovative design method, the cost of the launch-pad is reduced to one-tenth and the acoustic level is also reduced to one-tenth, which is the top level quietness in the world.

JAXA has been driving the incorporation of HPC technology into the design process of spacecraft. They are not mere troubleshooting tools any more. JAXA keeps on showing the future of the development process with HPC technology in aerospace field for 5 or 10 years ahead.

References

1. FUJITSU Supercomputer PRIMEHPC FX100. http://www.fujitsu.com/global/products/computing/servers/supercomputer/primehpc-fx100/
2. STREAM: Sustainable memory bandwidth in high performance computers. http://www.cs.virginia.edu/stream/
3. Takaki, R., Yamamoto, K., Yamane, T., Enomoto, S., Mukai, J.: The development of the UPACS CFD Environment. In: Veidenbaum, A., Joe, K., Amano, H., Aiso, H. (eds.) Lecture Notes in Computer Science, p. 2858. Springer (2003)
4. Morita, Y., Imoto, T., Tokudome, S., Ohtsuka, H.: Epsilon rocket launcher and future solid rocket technologies. In: 28th International Symposium on Space Technology and Science (2011)
5. Eldred, M., et al.: Acoustic loads generated by the propulsion system. NASA SP-8072 (1971)
6. Tsutsumi, S., Takaki, R., Nakanishi, Y., Okamoto, K., Teramoto, S.: Numerical study on acoustic generation of a supersonic jet impinging to deflectors. J. Acoust. Soc. Am. 134(5), 4057 (2013)

High Resolution Climate Projections Using the WRF Model on the HLRS

Viktoria Mohr, Thomas Schwitalla, Volker Wulfmeyer
and Kirsten Warrach-Sagi

Abstract Considering the projections of different climate scenarios, global mean surface temperature is expected to rise over the 21st century accompanied by an increase of other weather extremes due to the past anthropogenic emissions of greenhouse gases. As the warming of many land areas is higher than on the global average, the impact of future climate conditions needs to be estimated rather on a regional scale. Thus, climate projections of spatially high resolution simulations are required in combination with their uncertainties and robustness. For a selected area, these simulations are being performed within the framework of EURO-CORDEX. ReKliEs-De (Regional Climate Ensembles Germany) is a Project which complements these simulations by providing projections of model ensembles about the development of future climate and climate extremes for Germany. With the Weather Research and Forecasting (WRF) model and its land surface model NOAH we are performing simulations from 1950 to 2100 with 0.44° (~50 km) and 0.11° (~12 km) resolution on the CRAY XC 40 at the High Performance Computing Center Stuttgart (HLRS). Results of the simulations on the 0.44° grid for a period from 1971–2000 and as comparison for two different future scenarios from 2071–2099 show an increase of the average temperature of up to 2–4 °C in Europe with respect to the chosen emission scenario. However, seasonally the changes are much more diverse.

1 Introduction

The expected climate change within the next decades is predicted to be caused by an increase of the emissions of anthropogenic CO_2 and other greenhouse gases. These processes will have a huge influence on the future climate and consequently on the society. Although climate change will be noticeable on the global scale, the effects will be much more diverse and extreme regarding the regional scale.

V. Mohr (✉) · T. Schwitalla · V. Wulfmeyer · K. Warrach-Sagi
Institut für Physik und Meteorologie, Universität Hohenheim,
70599 Garbenstrasse 30, Stuttgart, Germany
e-mail: viktoria.mohr@uni-hohenheim.de

© Springer International Publishing AG 2016
M.M. Resch et al. (eds.), *Sustained Simulation Performance 2016*,
DOI 10.1007/978-3-319-46735-1_14

General circulation models (GCMs) currently are the most advanced tools for simulating the response of the global climate system to increasing greenhouse gas concentrations. They are able to numerically solve the equations of physics hence, GCMs have the ability of representing physical processes of the atmosphere, ocean, cryosphere and the land surface using a 3-D grid over the globe with a typical resolution. Nowadays, typically GCMs use a horizontal resolution of 100–200 km. However, to better understand also regional climate phenomena such as local extremes and in order to assess the effect of the expected climate change, scientists and end users like federal agencies and climate impact and adaptation researchers require projections on the regional scale with a higher horizontal resolution.

The Coupled Model Intercomparison Project Phase 5 (CMIP5) [1] provides a framework for coordinated climate change experiments, where 20 different modelling groups carried out global climate projections with their GCMs, contributing to the latest assessment report of the Intergovernmental Panel on climate change (IPCC). Those projections are based on Representative Concentration Pathways (RCPs) [2], representing four different possible greenhouse gas (GHG) concentration scenarios of the future climate. Namely, these scenarios are the RCP8.5, RCP6, RCP4.5 and RCP2.6 scenario, indicating the possible range in the change of radiative forcing (in W/m^2) by the year 2100 relative to pre-industrial values.

In 2009 the COordinated Regional climate Downscaling EXperiment (CORDEX, http://wcrp-cordex.ipsl.jussieu.fr) was established by the World Climate Research Programme (WCRP), in order to provide ensembles of regional climate simulations on a higher spatial resolution. The task within CORDEX is to apply the GCMs which contributed to the CMIP5 database for the boundary forcing of different regional climate models (RCMs). CORDEX is separated into different sub-groups, covering all continental regions of the globe, serving as input for climate change and adaptation studies. EURO-CORDEX, the european branch of the CORDEX initiative, is focusing on grid sizes of 0.11° (~12 km), but simulations using a resolution of 0.44° (~50 km) are carried out simultaneously.

The ReKliEs-De project funded by the Federal Ministry of Education and Research (BMBF), contributes to EURO-CORDEX by carrying out a certain number of regional climate projections. Several institutes in Germany are part of ReKliEs-De applying different downscaling techniques (dynamical and statistical downscaling), by forcing their RCMs with CMIP5 data. The goal of the ReKliEs-De project (http://www.reklies.hlnug.de/) is to provide robust informations about the future evolution of the climate in Germany based on regional climate projections using a high spatial resolution of 0.11°. Beside scientifically analysing the results of the simulations, the focus is also laid on producing user friendly data by providing special climate indices which can be applied easily by the end users. Today, the dynamical downscaling of GCM data is a well respected standard technique to produce high resolution climate projections.

An ensemble of RCMs participating in EURO-CORDEX, forced with ERA-Interim reanalyses data [3] were evaluated e.g. by [4]. Although model biases, predominantly cold and wet biases, emerged in the majority of the experiments, the authors documented the general ability of the RCMs to represent the basic spatiotemporal patterns for the european climate.

The scope of this part of the WRFCLIM project at HLRS (see also [5]), is to produce high resolution regional climate projections in the framework of ReKliEs-De on 0.11°. We will carry out this long term simulations by dynamically downscaling different GCM model output from 1958 to 2100 with the Weather Research and Forecasting (WRF) [6] Model.

In the following chapter a technical description to the simulations within WRF-CLIM at HLRS is presented. A summary of the current analysis and some preliminary results of the simulations which were carried out from April 2015 to April 2016 are reported as well.

2   Simulation Setup with WRF and Forcing Data

The climate simulations are carried out on the CRAY XC40 System at the HLRS. The downscaling process of the coarse GCM output data to the required 0.44° and 0.11° grid resolution is performed by the WRF model version 3.6.1 with its coupled land surface model NOAH [7]. WRF uses the dynamical technique by numerically solving the governing equations of the atmosphere on a finer grid. The simulations are forced 6 hourly at the lateral boundaries with the sea surface temperature and the 3-D fields of temperature, horizontal winds and moisture from the forcing GCMs. We apply a one-way nesting approach via 0.44° to 0.11°. WRF was compiled at HLRS with PGI 14.7 and applied in a hybrid configuration using MPI and OpenMP to optimize the speed of the simulation. The model configuration was set to the same parameterizations like it was applied in [8]. Table 1 shows some technical details of the planned and the so far partly finalized simulations performed on hazelhen.

The WRF projections presented in this chapter are carried out within the frame of EURO-CORDEX to create climate projections. The simulation domains which were specified within CORDEX are displayed in Fig. 1 (left). The analysis focuses on Germany and its contributing river catchments area as shown in Fig. 1 (right). For the future climate projections, four different GCMs and two different RCP scenarios of the CMIP5 project, are applied as boundary forcing with the WRF model. The historical runs of the GCMs cover the period from 1850 to 2005. This period is forced by observed atmospheric composition changes of anthropogenic and natural sources. The RCP scenarios of the GCMs cover the period from 2006 to 2100. They represent mitigation scenarios that assume policy actions will be taken into account to achieve certain emission targets [1]. The numbers of the RCPs give a rough estimate of the range in the change of the radiative forcing by the year 2100 relative to the preindustrial values. The forcing data we applied, the resolution of the GCMs, its scenarios and the chosen simulation period is presented in Table 2.

Table 1 Technical details of WRF simulations. Note: the raw model output size is minimized after simulation in the postprocessing process as the amount of the results cannot be stored in its raw output

Simulation	Nr. CPUs with openMPI	Simulation period	Nr. of grid cells	Δt (s)	Walltime (h)	Nr. of simula- tions	Raw output size
0.44° historical	1536	01.01.1958– 31.12.2005	129 × 139 × 50	180	300	4	42 TB
0.44° RCP8.5/2.6	1536	01.01.2006– 31.12.2100	129 × 139 × 50	180	570	5	99 TB
0.11° historical	5400	01.01.1958– 31.12.2005	452 × 460 × 50	60	1200	4	700 TB
0.11° RCP8.5/2.6	5400	01.01.2006– 31.12.2100	452 × 460 × 50	60	2280	5	1330 TB

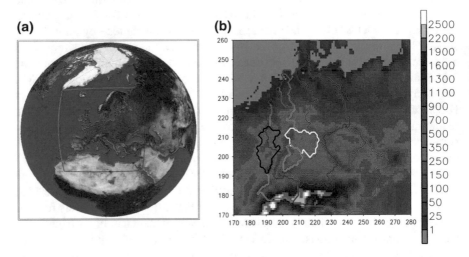

Fig. 1 Simulation domain as specified by CORDEX (**a**), and focal domain within ReKliEs-De with Germany (*red*) and the areas of its main river catchments (*coloured*) (**b**)

3   Results

Based on the WRF downscaling simulations to the 50 km grid, first results of the different forcing GCMs are presented (see GCM description in Table 2). Simulations were carried out for the EURO CORDEX domain with lateral boundaries from approximately 25°N to 75°N and 30°W to 50°E. Since our model simulations are still ongoing by this time (May 2016) exemplary results of the 2 m temperature will

Table 2 Applied GCM data for the WRF simulations

GCM	scenarios	Simulation Period	GCM resolution
MPI-ESM-LR	historical	1958–2005	$1.8653° \times 1.875°$
	RCP8.5	2006–2100	
	RCP2.6	2006–2100	
MIROC5	historical	1958–2005	$1.4008° \times 1.40625°$
	RCP8.5	2006–2100	
HadGEM2-ES	historical	1958–2005	$1.25° \times 1.875°$
	RCP8.5	2006–2100	
EC-EARTH	historical	1958–2005	$1.1215° \times 1.125°$
	RCP8.5	2006–2100	

be presented. In the first part the focus is based on a time period of the historical simulations from 1971–2000 represented by the four forcig GCMs which were applied
in our study. The second part shows a time period of the projected simulations for the
two future scenarios RCP2.6 and RCP8.5 represented by the MPI-ESM-LR model
for the years 2071–2099. The average near surface temperature is presented in the
following based on the annual and on the seasonal scale for the Northern Hemisphere
(NH) summer and winter. It is emphasized that the results of this study are based on
just one regional model and will need further analyses within the EURO-CORDEX
ensemble and the ReKliEs-De project.

3.1 Historical Temperatures

The simulated annual average near surface temperature from 1971–2000 is given in
Fig. 2. The four simulations show a common agreement of the average temperature
pattern for Europe, which indicates a generally good performance of the models. The
temperature ranges from around 264 K (-9 °C) in the North to 294 K (21 °C) in the
Southern part of the simulation domain. However, some differences in the tempera-
ture range is apparent among the GCMs. Compared to HadGEM2-ESM Fig. 2b and
MPI-ESM-LR Fig. 2d, lower temperatures are simulated by MIROC5 Fig. 2c and
EC-EARTH Fig. 2a in the Northern and in the Eastern part of the CORDEX domain.
Most of the GCMs show a cold bias. MIROC5 indicated to have the strongest bias
compared to the other three which are presented here.

The average near surface temperature for the time period 1971–2000 as simulated
by WRF with the four different forcing GCMs is given for the winter season (DJF)
in Fig. 3 and for the summer season (JJA) in Fig. 4. The average temperatures during
NH winter vary from around 258 K (-15 °C) in North-East to around 290 K (17 °C)
over the atlantic ocean in the south-west of the domain. During NH summer the
lowest temperatures occur over Iceland in the north-east and over the arctic ocean

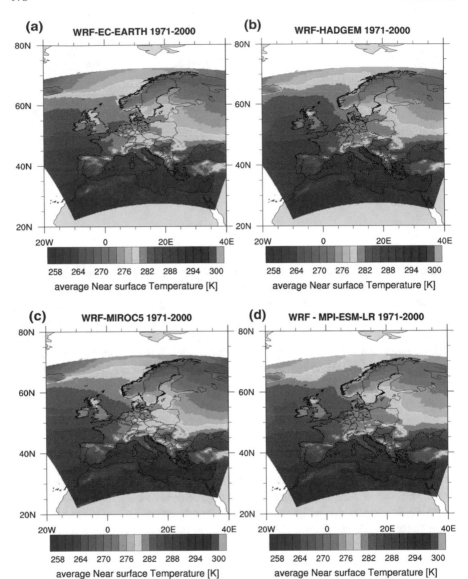

Fig. 2 Average near surface temperature on the 50 km resolution simulations from WRF. Boundary forcing with the different GCMs: EC-EARTH (**a**), HadGEM2-ESM (**b**), MIROC5 (**c**) and MPI-ESM-LR (**d**)

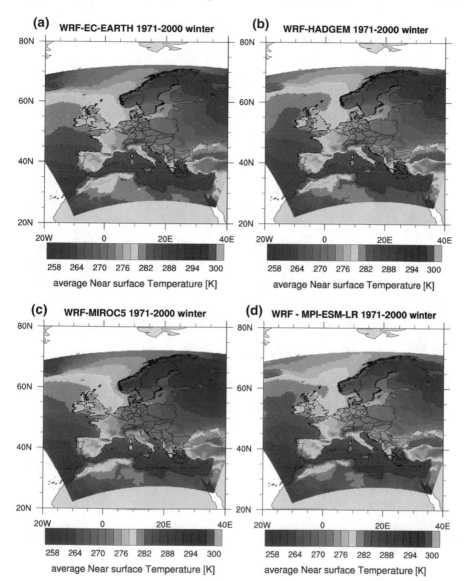

Fig. 3 Average near surface temperature on the 50 km resolution simulations from WRF during winter. Boundary forcing with the different GCMs: EC-EARTH (**a**), HadGEM2-ESM (**b**), MIROC5 (**c**) and MPI-ESM-LR (**d**)

180   V. Mohr et al.

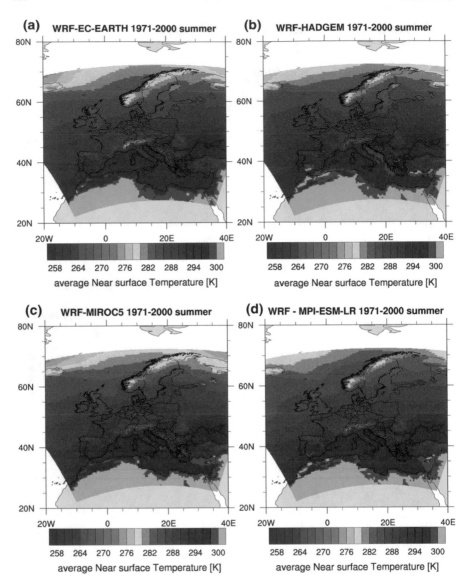

Fig. 4 Average near surface temperature on the 50 km resolution simulations from WRF during summer. Boundary forcing with the different GCMs: EC-EARTH (**a**), HadGEM2-ESM (**b**), MIROC5 (**c**) and MPI-ESM-LR (**d**)

with 276 K (3 °C). Highest temperatures in Europe aside from the Sahara desert are around 296 K (23 °C) over the mediteranean and the greek islands. Similar as for the annual average, EC-EARTH and MIROC5 indicate lower temperatures (Figs. 3a–c and 4a–c) in both presented seasons compared to the two GCMs HadGEM2-ESM and MPI-ESM-LR (Figs. 3b–d and 4b–d).

3.2 Projections for Temperature Changes

The future changes, based on the differences of the simulated average temperatures from 1971–2000 and 2071–2099, is shown in Fig. 5. Following the RCP2.6 scenario, the highest temperature increase reaches 2–2.4 °C above the Norwegian Sea and the Russian peninsula Kola. Following the RCP8.5 scenario, highest temperature changes are simulated above the Russian peninsula with an increase of around 5 °C.

The seasonal changes are displayed in Fig. 6 for the winter and for the summer averages. During winter the RCP2.6 scenario indicates a maximum increase of the temperatures over the peninsula of Russia with up to 4.5 °C (Fig. 6a). For the RCP8.5 scenario the temperature difference is even higher, revealing an increase of up to 6 °C (Fig. 6b). Beside Russia, also the eastern part of Turkey will experience as simulated for this high emission scenario a high temperature increase of around 5 °C compared to the historical period.

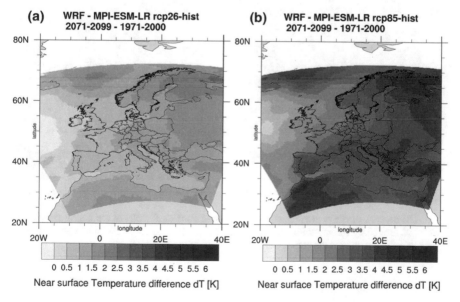

Fig. 5 Differences of the average near surface temperature on the 50 km resolution simulations from WRF. Boundary forcing with MPI-ESM-LR for 1971–2100 RCP 26 (**a**) and RCP8.5 (**b**)

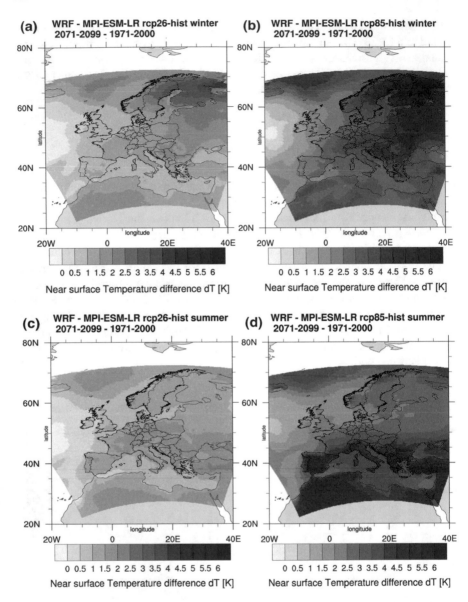

Fig. 6 Differences of the average near surface temperature on the 50 km resolution simulations from WRF. Boundary forcing with MPI-ESM-LR for 1971–2100 for the winter season: RCP 26 (**a**) and RCP8.5 (**b**) and for the summer season RCP 26 (**c**) and RCP8.5 (**d**)

Compared to the temperature changes in winter, the increase of the average summer temperature is less intense for both scenarios. The increase in the temperature will reach around 1.5 °C in central Europe and in the southern part of the norwegian sea for the RCP2.6 scenario (Fig. 6a). For the RCP8.5 scenario the highest temperature increases are simulated over south and south eastern Europe and northern africa indicating a temperature increase of around 5 °C in Spain and up to 6 °C in Turkey.

4   Conclusion

The results shown in this chapter highlighted preliminary results based on the near surface temperature of the downscaling of four GCMs (EC-EARTH, HadGEM2-ESM, MIROC5 and MPI-ESM-LR) from the original model grid of around 150 km with WRF to a refined grid on 50 km resolution for the EURO-CORDEX domain. Further downscaling simulations to the 12 km grid for the historical and future time periods are still ongoing. It is expected to finalize the simulations as described in Table 1 by the end of this year (2016) accompanied with further improvements of the projection skills due to a better representation of the orographic effects.

The short insight of the seasonal differences which was shown in this chapter, pointed out the need of further analyses. In particular different variables and climate indices need to be investigated in more detail as the information about future changes of the climate is a major need for cultivators world wide. Precipitation is a parameter which needs to be pointed out and analysed carefully as it showed to be highly variable throughout the year in germany and also Europe.

We would like to point out the importance of analysing a model ensemble rather than choosing a single RCM when predictions of the evolution of climate parameters are being done. To achieve robust predictions, ReKliEs-De is an appropriate platform. Within this project a huge ensemble member of climate simulations can be realized with the contribution of several RCMs applying different techniques and GCMs.

Acknowledgements This work is part of the ReKliEs-De project funded by the BMBF (Federal Ministry for Education and Research) and the Research Unit 1695 funded by the DFG (Deutsche Forschungsgemeinschaft). We would like to thank the staff for the support of the DKRZ (Deutsches Klimarechenzentrum) to give access to the GCM data. Computational Resources for the model simulations on the HLRS CRAY XC40 within WRFCLIM were kindly provided by HLRS, we appreciate the great support.

References

1. Karl, E.: Taylor, Ronald J Stouffer, and Gerald A Meehl. An overview of cmip5 and the experiment design. Bulletin of the American Meteorological Society **93**(4), 485–498 (2012)
2. Van Vuuren, Detlef P., Edmonds, Jae, Kainuma, Mikiko, Riahi, Keywan, Thomson, Allison, Hibbard, Kathy, Hurtt, George C., Kram, Tom, Krey, Volker, Lamarque, Jean-Francois, et al.: The representative concentration pathways: an overview. Climatic change **109**, 5–31 (2011)

3. Dee, D.P., Uppala, S.M., Simmons, A.J., Paul Berrisford, P., Poli, S.Kobayashi, Andrae, U., Balmaseda, M.A., Balsamo, G., Bauer, P., et al.: The era-interim reanalysis: Configuration and performance of the data assimilation system. Quarterly Journal of the Royal Meteorological Society **137**(656), 553–597 (2011)
4. Kotlarski, Sven, Keuler, Klaus: Ole Bossing Christensen, Augustin Colette, Michel Déqué, Andreas Gobiet, Klaus Goergen, Daniela Jacob, Daniel Lüthi, Erik van Meijgaard, Gregory Nikulin, Christoph Schär, Class Teichmann, Robert Vautard, Kirsten Warrach-Sagi, and Volker Wulfmeyer. Regional climate modeling on european scales: a joint standard evaluation of the euro-cordex rcm ensemble. Geoscientific Model. Development **7**(4), 1297–1333 (2014)
5. Kirsten Warrach-Sagi, Thomas Schwitalla, Hans-Stefan Bauer, et al. A regional climate model simulation for euro-cordex with the wrf model. In *Sustained Simulation Performance 2013*, pages 147–157. Springer, 2013
6. William C Skamarock, Joseph B Klemp, Jimy Dudhia, David O Gill, Dale M Barker, Wei Wang, and Jordan G Powers. A description of the advanced research wrf version 2. Technical report, DTIC Document, 2005
7. Chen, Fei, Dudhia, Jimy: Coupling an advanced land surface-hydrology model with the penn state-ncar mm5 modeling system. part i: Model implementation and sensitivity. Monthly Weather Review **129**(4), 569–585 (2001)
8. Warrach-Sagi, Kirsten, Schwitalla, Thomas, Wulfmeyer, Volker, Bauer, Hans-Stefan: Evaluation of a climate simulation in europe based on the wrf-noah model system: precipitation in germany. Climate Dynamics **41**(3–4), 755–774 (2013)

Towards Aerodynamic Characteristics Investigation Based on Cartesian Methods for Low-Reynolds Number Flow Simulation

Daisuke Sasaki, Yuya Kojima, Daiki Iioka, Ryohei Serizawa and Shun Takahashi

Abstract Micro Aerial Vehicles (MAVs) are recently focused for various usage such as monitoring, photographing, and filming. One of the issues of MAVs is the limitation of operation time. An efficient configuration is required for MAVs, however, complex low-Reynolds number flows causes the difficulty. In this research, Cartesian-based CFD approach is applied to a flat plate, a NACA0012 airfoil, and a circular arc at low-Reynolds number flows to investigate the aerodynamic characteristics. Block-structured Cartesian mesh solver, Building-Cube Method, was capable to investigate the complicated flowfields at lower angles of attacks.

1 Introduction

Micro Air vehicles (MAVs) have been widely researched and developed in recent years for various purposes, such as the disaster monitoring and aerial photographs. The aerodynamic characteristics of MAVs is different from large passenger aircrafts because of low Reynolds number flows. Conventional airfoils such as NACA series, which are thick and streamlined shapes, generally produce low aerodynamic performance at low Reynolds number in MAVs flight regime. Instead, it is well-known that thin airfoils can often increase the aerodynamic performance by making use of laminar separation bubble at sharp leading-edge [1]. Therefore, the complicated flow phenomena needs to be precisely investigated through Computational Fluid Dynamics (CFD) and wind/water tunnel tests to improve aerodynamic performances [2, 3].

It is required to solve complicated flowfields precisely to obtain accurate aerodynamic performance at low Reynolds number conditions. Cartesian mesh is expected to solve such flowfields because it can prevent numerical vortex dissipation and maintain vortices. Building-Cube Method (BCM [4]) is a block-structured Cartesian mesh

D. Sasaki (✉) · Y. Kojima · D. Iioka
Kanazawa Institute of Technology, 3-1 Yatsukaho, Hakusan, Japan
e-mail: dsasaki@neptune.kanazawa-it.ac.jp

R. Serizawa · S. Takahashi
Tokai University, 4-1-1 Kitakaname, Hiratsuka, Japan

© Springer International Publishing AG 2016
M.M. Resch et al. (eds.), *Sustained Simulation Performance 2016*,
DOI 10.1007/978-3-319-46735-1_15

CFD solver proposed for efficient parallel computation. The objective of the study is to conduct incompressible flow computations on various airfoils at different computational conditions to investigate the capabilities of BCM solver for low-Reynolds number flows. The computational models are a flat plate, a thick airfoil (NACA0012), and a circular arc, all of which have different flow characteristics.

2 Building-Cube Method

This study adopts block-structured Cartesian mesh solver, Building-Cube Method (BCM). The code employs incompressible Navier-Stokes equations. BCM simplifies pre-process, computation, and post-process. It is easy to generate mesh for a real-world complicated shape, and also to implement spatial higher-order accuracy scheme. BCM divides computational domain with many blocks, named Cubes, which are shown in Fig. 1. Equally-spaced Cartesian mesh, so-called Cells, are then filled in each Cube. Computational domain is composed of many Cubes with different size, but each Cube possess the same number of Cells regardless the Cube size. The method allocates a lot of small-size Cubes near the model where physical quantities change largely, thus dense mesh is only distributed to the vicinity of the model. Therefore, in the domain allocated with minimum cell, high spatial accuracy is maintained and also numerical vortex dissipation is prevented. This is advantage against complicated flowfields with separation bubble and large vortices around thin airfoil under low Reynolds number range. In addition, it can be easy to conduct parallel computation while maintaining parallel efficiency. Individual Cubes are computed independently, thus BCM needs to exchange physical quantities between the adjacent

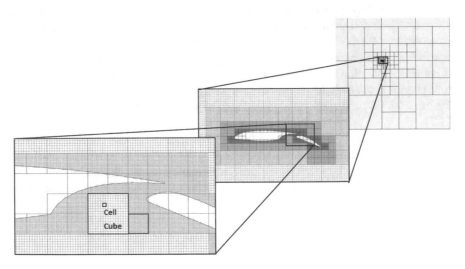

Fig. 1 Cube allocation around multi-element airfoil (cube and cell)

Fig. 2 Flowchart of
fractional step method

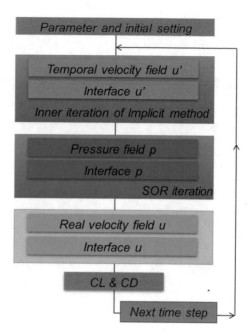

Cubes during the computational process. In this computation, there are overlap area
of 2 Cells between adjacent Cubes. Therefore, at the exchange process, Cubes of the
same size can maintain the interpolation accuracy at Cube boundaries. However, in
the case of Cubes of different size between adjacent Cubes, linear interpolation is
conducted from small size Cube to large size one. Nearest neighbor interpolation is
conducted from large size to small size Cubes.

The governing equations are equation of continuity and two- and three-dimensional Navier-Stokes equations. It is discretized on the collocated-mesh scheme.
Pressure and velocity are located at cell-center, and additional quantity called con-
travariant velocity is located at cell-face. Navier-Stokes equation is integrated in time
by fractional step method in Fig. 2. Adams-Bashforth explicit scheme is employed
for convective and diffusive terms. The convective term is discretized using first-
order accurate upwind difference scheme. Second-order accurate central difference
scheme of second derivative is adopted for the diffusive term. Wall is treated as staircase representation. In this study, turbulent model is employed, and thus the inflow
boundary condition is set as laminar flow.

3 Computational Model and Conditions

The target of the computation is 2 % flat plate, NACA0012 airfoil, and circular arc
as shown in Fig. 3, which were experimented by Kuroda et al. [2]. The mesh near the
airfoil is also shown in the figure. The characteristic length is chord length c, and the

Fig. 3 Computational model: flat plate *(left)*, NACA0012 *(middle)*, circular arc *(right)*

Fig. 4 Computational
domain

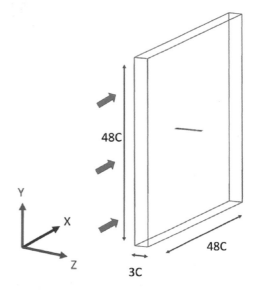

Table 1 Computational conditions

Item	Flat plate	NACA0012	Circular arc
Reynolds number	5.0×10^3	2.8×10^3	2.8×10^3
Min. cell size	5.86×10^{-3}	2.93×10^{-3}	2.93×10^{-3}
Cells in a cube	16^3	16^3	16^3
Number of cubes	3,112	5,112	5,112
Number of cells	12,746,572	20,938,752	20,938,752

span length is three times of the chord length as shown in Fig. 4. Three-dimensional flow analysis is conducted and the periodic boundary condition is applied to both side of the wing tips. The inflow of computational domain is laminar flow. Table 1 describes computational condition.

4   Results and Discussion

4.1   *Aerodynamic Computations of a Flat Plate*

Two- and three-dimensional BCM incompressible solvers were applied to flows around a flat plate. The aerodynamic coefficients of the computations are plotted

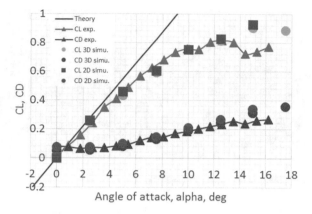

Fig. 5 Aerodynamic coefficient of a flat plate

in Fig. 5. Lift coefficients of two-dimensional and three-dimensional computations are well-matched with experiments except for the very high angles of attack, and the non-linearity of the C_L distribution are captured. However, drag coefficients show the discrepancy between two-dimensional and three-dimensional results. Two-dimensional results are relatively matched with experiments at lower angles of attack, while three-dimensional results are matched with experiments at large angles of attack. Freestream velocity distributions of 2.5 and 10° are shown in Figs. 6 and 7, respectively. At the low angle of attack (2.5°), the distributions are almost identical in

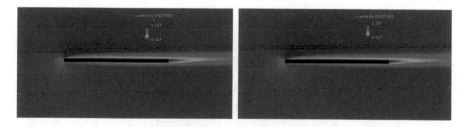

Fig. 6 Freestream velocity contours at 2.5°: 3D results (*left*), 2D results (*right*)

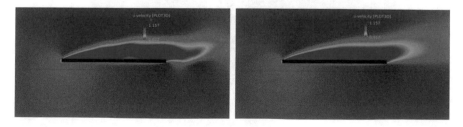

Fig. 7 Freestream velocity contours at 10°: 3D results (*left*), 2D results (*right*)

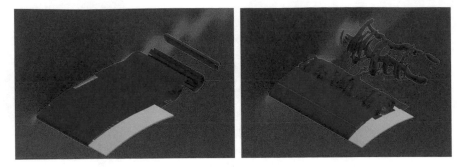

Fig. 8 Vortex structures (iso-surface).: 2° (*left*), 10° (*right*)

two-dimensional and three-dimensional cases. However, the difference is observed at the wake region in Fig. 7. Figure 8 shows the vortex structure at 2 and 10°. It is obvious that the generated vortices are three-dimensional at higher angles of attack, thus three-dimensional computations agree well with experiments.

4.2 Aerodynamic Computations of NACA0012

Figure 9 shows the computational and experimental aerodynamic coefficients of NACA0012. Experimental results show the slight non-linearity of C_L distribution, while computational results show the linear relation with regard to angles of attacks.

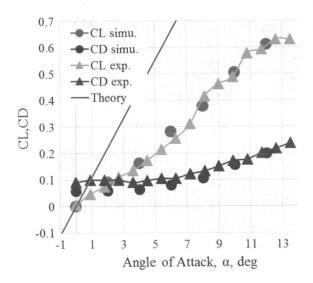

Fig. 9 Aerodynamic coefficient of a NACA0012 airfoil

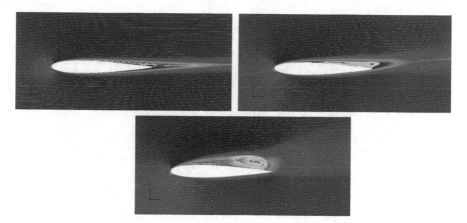

Fig. 10 Freestream velocity contours around NACA0012 airfoil.: 2° (*top left*), 6° (*top right*), 10° (*bottom*)

Freestream velocity contours with streamlines of NACA0012 airfoil are shown in Fig. 10. The separation location at 2° is close to the middle and the small separation vortex is formed. As the angle of attacks increases, the separation location moves forward and the separation vortex at the trailing edge becomes larger as appeared at 6°. The further increases of angle of attacks causes the separation location forward and the large separation vortex formed. Since the lift slope becomes lower at high angles of attacks, it is expected that the movement of separation location influences the non-linearity of C_L distribution.

4.3 Aerodynamic Computations of Circular Arc

Figure 11 shows the computed aerodynamic coefficients of a circular arc. Experimental results show the linear distribution of C_L, and computational results also show the similar trend that the linear relation with regard to angles of attacks is observed at lower angles of attacks. However, large discrepancy of computed C_L with experiments is observed at higher angles of attacks. Freestream velocity contours and streamlines in Fig. 12 shows that the separation occurs at near the maximum camber location at 2 and 6°, while the separation occurs at near the leading edge and the upper surface is covered with large vortices in computations. The strong unsteadiness is observed at 2°, but the unsteadiness gets weaker as the increase of angles of attacks to 6°. At the lower angles of attacks, the vortex is formed at near the trailing edge. On the other hand, separation location moves forward and the vortex covers the whole upper surface at higher angles of attacks, which leads the rapid increase of lift coefficient. It is possible the present computation at 10° or higher can not fully capture the flow features due to the lack of wake-region mesh resolution or the staircase

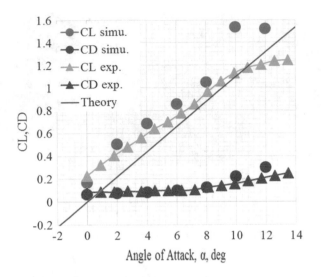

Fig. 11 Aerodynamic coefficient of a circular arc1

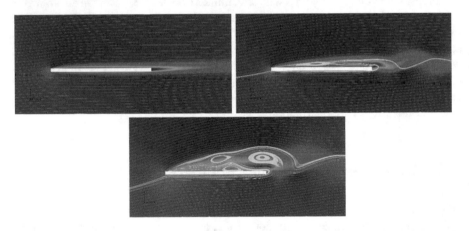

Fig. 12 Freestream velocity contours around a circular arc.: 2° (*top left*), 6° (*top right*), 10° (*bottom*)

representation of curved surface or the employment of 1st order accurate scheme., which causes the difference with experimental values. Thus, further investigations of flowfields will be needed at higher angles of attacks.

5 Concluding Remarks

Cartesian-mesh based CFD solver, Building-Cube Method, was applied to various airfoils at low Reynolds number flows to investigate the aerodynamic characteristics. Three-dimensional flow computations are needed, where three-dimensional effects of flow vortices can not be negligible. A conventional thick airfoil such as NACA0012 can not achieve good aerodynamic performance at the flow regime because of the separation vortex, while thin airfoil such as circular arc can produce higher aerodynamic performance because of the vortex formed. Thus, the present flow solver is capable to investigate the complicated flowfields at low Reynolds number region. However, more precise investigations are needed in terms of schemes and mesh density for high angles of attacks.

Acknowledgements Part of the work was carried out under the Collaborative Research Project of the Institute of Fluid Science, Tohoku University. Part of this research used computational resources of the HPCI system provided by Cyberscience Center at Tohoku University through the HPCI System Research Project (Project ID:hp140138 and hp150130).

References

1. Alam, M., Sandham, N.D.: Direct numerical simulation of short laminar separation bubbles with turbulent reattachment. J. Fluid Mech. **410**, 1–28 (2000)
2. Kuroda, T., Okamoto M.: Unsteady aerodynamic forces measurements on oscillating airfoils with heaving and feathering motions at very low Reynolds Number. In: Proceedings Asia-Pacific International Symposium on Aerospace Technology 2013. Takamatsu (2013)
3. Iioka, D., Kojima, Y., Okamoto, M., Sasaki, D., Obayashi, S., Shimoyama, K.: Analysis of thin angular airfoils using block-structured Cartesian Mesh CFD. In: Prooceedings Asia-Pacific International Symposium on Aerospace Technology 2015, (2015)
4. Sakai, R., Obayashi, S., Matsuo, K., Nakahashi, K.: Practical large-scale turbulent flow simulation using building-cube method. In: Proceedings 45th Fluid Dynamics Conference/Aerospace Numerical Simulation Symposium 2013, JSASS-2013-2116-A, Funahori (2013)

Printed in the United States
By Bookmasters